BORNEO LOG

WILLIAM W. BEVIS

UNIVERSITY OF WASHINGTON PRESS

Seattle & London

BORNEO
LOG

THE STRUGGLE FOR SARAWAK'S FORESTS

Borneo Log: The Struggle for Sarawak's Forests is winner of the 1995 Western States Book Award for Creative Nonfiction. The Western States Book Awards are a project of the Western States Arts Federation. The awards are supported by Crane Duplicating Services, The Elaine Horwich Memorial Cultural Fund (Arnold Horwitch, Trustee), and the Witter Bynner Foundation for Poetry. Additional funding is provided by the National Endowment for the Arts.

Library of Congress Cataloging-in-Publication Data
Bevis, William W., 1941–
    Borneo log : the struggle for Sarawak's forests / William W. Bevis.
    p.   cm.
    Includes bibliographical references (p.   ).
    ISBN 0–295–97416–8 (alk. paper)
    1. Logging—Malaysia—Sarawak.   2. Deforestation—Malaysia—Sarawak.   3. Rain forest conservation—Malaysia—Sarawak.   4. Plywood industry—Japan.   5. Bevis, William W., 1941—Journeys—Malaysia—Sarawak.   6. Sarawak—Description and travel. I. Title.

SD538.3.M4B48 1995	95–18317
959.5'4—dc20	CIP

The paper used in this publication meets the minimum requirements of American National Standard for Information Sciences—Permanence of Paper for Printed Library Materials, ANSI Z39.48-1984.

To my wife, Juliette Taft Crump

CONTENTS

ACKNOWLEDGMENTS

First I would like to thank the Kayan, Kenyah, and Penan upriver who helped in so many ways. Publishing names might cause problems, but day after day the people of the Baram were courteous, informative, and honest.

A number of ministers and public officials also spoke freely and at times off the record; they too must be thanked anonymously.

On the record, several public figures whose views are well known helped a great deal: Datuk James Wong was generous with his time and energy. Harrison Ngau, Sim Kwang Yang, Joseph Tingang, Thomas Jalong, and Chee Yoke Ling gave me invaluable advice and information. In addition, Bruno Manser was exceptionally kind, sending biographical details and pages from his journal even as he prepared his own book on Sarawak, published in German.

At the manuscript stage, many friends in Missoula offered editing advice: writers Peter Stark and Bryan Di Salvatore gave me incisive, constructive, professional readings — and every author knows the value of that. Lu Haas — Junior and Senior — gave welcome advice. Colleagues Jill Belsky, Bill Kittredge, and Paul Svrcek also helped, and useful comments and encouragement came from colleagues in Asian Studies at the Mansfield Center at the University of Montana, including John Spores, Phil West, and Dennis O'Donnell. Thanks, all.

Throughout this project, several organizations and their staff members assisted me in various ways: Rainforest Action Network in San Francisco, Cultural Survival in Boston (perhaps the only archive of the Utusan Konsumer in the United States), JATAN and Yoichi Kuroda in Tokyo, Judith Meyer and the Berkeley-Borneo Big Home Project, and Mike Mease at Cold Mountains, Cold Rivers in Missoula, who produced a video, "SOS Sarawak", using some of my footage and information. The Malaysian government sponsored an excellent conference (one of a series) at Ohio University on "Malaysia and the Environment," which represented the government position well, and which also respected dissenting voices. A sabbatical leave from the Univer-

sity of Montana made my travel possible, and the book was nurtured
by the kind and efficient folks at the University of Washington Press.

Finally, I want to thank everyone at SAM in Marudi and Penang,
and SAM's benefactors in the Consumers Association of Penang, for
without their primary work in the field—investigating, organizing,
publishing—almost nothing would be known of the events upriver in
Sarawak. A few people, in a small organization, can make a big difference. We could use their help now in our own Northwest.

Part of the proceeds from the sale of this book go to the SAM office
in Marudi. The author recommends support of Friends of the Earth
as the parent of SAM, and support of a number of other organizations
long involved in Sarawak:
— Consumers Association of Penang (CAP)
  87 Cantonment Road
  10250 Penang, Malaysia
— Cultural Survival
  215 First Street
  Cambridge, Massachusetts 02142
— Friends of the Earth
  1025 Vermont Avenue, N.W.
  Washington, D.C. 20005
— Japan Tropical Forest Action Network (JATAN)
  801 Shibuya Mansion, 7-3-1
  Uguisudani-cho, Shibuya, Tokyo 150, Japan
— Rainforest Action Network (RAN)
  450 Sansome Street, Suite 700
  San Francisco, California 94111
— Society for Threatened People
  Eigenstrasse 15, CH-8008
  Zurich, Switzerland
— Sahabat Alam Malaysia (SAM)
  43 Salween Road
  10050 Penang, Malaysia

BORNEO LOG

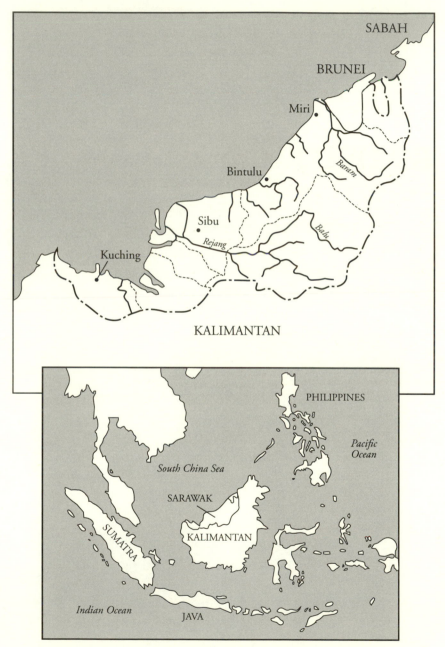

Map 1. Sarawak (adapted from Evelyne Hong, *Natives of Sarawak*)

"Sarawak is the belly of the beast in a couple of different ways. It's the oldest rain forest on the planet. Its situation is so desperate that if we can turn it around we can perhaps save a significant part of what's left of the world's rain forests. At some point, you've got to draw your line in the sand and say 'Don't cross this line.' Borneo is that line in the sand." — Randy Hayes, Rainforest Action Network, quoted in *The New Yorker,* May 27, 1991

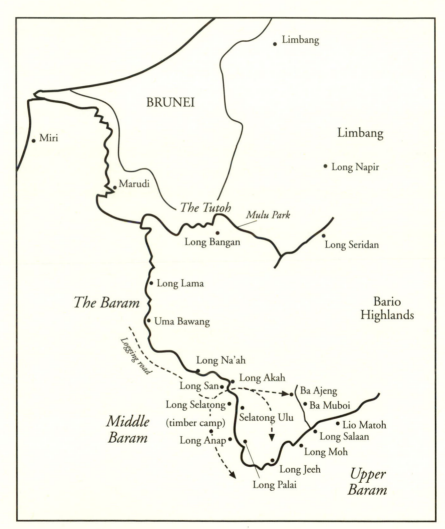

Map 2. Travel route along the Baram River

PROLOGUE

What can I tell you that you do not know? That as I sit here, home in Montana after two years away, Borneo is a dream of green, of insects honking like car horns, of rain like cataracts, of beautiful people who put a hand into yours as if placing the world's last butterfly, fallen and dazed, back on a leaf. It rests there; they do not shake. That these people once hunted heads, and so were famous; that headhunting is irrelevant now and will not be mentioned again; that their land is being taken away, bit by bit, the length and breadth of Sarawak, north Borneo; that their rainforest, the oldest in the world, is being cut; that they stand in feathered hats, earlobes stretched to the breasts, spears and blowpipes in hand, in front of Japanese bulldozers.

May I tell you that right now on foot-polished hardwood porches far up the brown rivers of Borneo, timber traders from the richest of countries are sitting down face to face with some of the last traditional tribal people on earth; that we, the consumers of this earth, are tearing the trees and vines from around their houses; that they do not want this; that at this moment one timber camp in the Baram is taking out $3.6 million a month of timber and paying for that privilege less than one percent, or $1.35 to each native of that district; that most of the timber goes to Japan, much of it to plywood forms for concrete which are thrown away after two uses; that the forest will be gone in seven to ten years. That the government and companies have no thought for tomorrow in Sarawak, although Mitsubishi is looking now toward Chile and New Zealand, and in 1990 bought extensive concessions in Alberta, Canada. As I sat in the Timber Section conference room in the Mitsubishi Soshi Annex, seventh floor, Ginza, Tokyo, March 5, 1991, they were considering the purchase of a Montana plywood mill near my home. "A man called yesterday," said the head of Mitsubishi Timber. "He wants eight million dollars for the mill." I guessed: "It's nothing." "It's nothing," he agreed. I felt for a second how the weak feel about the strong, how Japan and most of the world had felt about America for two generations, how Borneo feels now about Japan. It is as much fun as looking at the head of Medusa, as sitting down across the table from your fate.

I have no wish to break your heart. Although many of the problems on earth cannot be solved, this one can be. The natives of Borneo might regain their land rights; Japan, under enough pressure, might stop the imports. Unlike Brazil and Africa, Sarawak is not plagued by millions of landless poor following the tractors into the forest. But the change had better come within three years, maybe five.

Borneo, the third largest island on earth, lies below Vietnam in the South China Sea. In the north of the island, the state of Sarawak was carved out of the jungle by Malay sultans and, after 1841, by an English adventurer named James Brooke, who became the improbable Rajah Brooke. He and his descendants ruled Sarawak for a century, until 1946. Soon after independence, Sarawak joined the new Malaysia. Sarawak, about the size of New York State, has 1.7 million people, approximately 45 percent native, 30 percent Chinese, and 20 percent Malay. Sarawak and neighboring Sabah, although now states of Malaysia, are somewhat autonomous. Coming from Malaysia you will have your passport checked—especially if you look like an environmental journalist.

There are only a few sections of paved road in Sarawak, mostly along the coast. Interior Sarawak is a world of jungle, brown rivers, dugout canoes, and outboards. Most of the natives live in longhouses along the two major rivers, the Baram and the Rajang, which are still the main routes of transportation.

A longhouse on the river is the traditional tribal building. Picture a long row of ranch houses side by side, wall to wall, a single-story building with twenty or thirty front doors the length of a football field. Now imagine it all up on stilts head high and add a front porch, roofed, also on stilts, running from one end of the building to the other. Each family's private door opens onto the huge porch. Rain or shine, this porch of clean hardwood, breezy and roofed, eight feet above the chickens and dogs and dirt, is a childrens' paradise. These settlements with their outbuildings (rice granaries on stilts, storage sheds, sometimes schools or a mission), fruit trees, gardens and flowers, belong to various tribes. On the Baram River, they are mainly Kayan tribe below, Kenyah in the middle and upper Baram, and Penan on the tributaries or "deep inside," as the natives say, in the jungle.

After a year on exchange at a university in Tokyo, my wife and I had the luck to spend six months in and out of Sarawak. These stories of Sarawak and Japan use the actual places, dates, and names from our travels unless otherwise noted. The exception is Richard, who is a composite of three different young men. Identifiable boys would not easily work again in the timber business. Similarly, information or events originating in one longhouse have sometimes been moved to another, to thwart reprisal.

Like most visitors, I first went upriver thinking of natives, nature, and adventure. I came out thinking of native land rights, government misadventures, and the nature of international trade. Meanwhile, in the background of my memory, cicadas sing and women are weaving mats in the deep shade of the porch. And now, sitting in Montana, I remember that by 1845 the beaver in Montana had all been trapped out for the European hat trade, and the Indians were shrunken by smallpox. And by 1885, the buffalo, which gave the Indians everything, were gone to make room for white ranchers speculating in beef, selling dead stock for their livestock on European exchanges. "When the buffalo went away," said the Crow Chief Plenty-coups, "the hearts of my people fell to the ground, and they could not lift them up again." It is happening again. The trees are the buffalo.

This is a book not so much *about* the native resistance to the logging as a series of stories from *inside* that struggle, from Tokyo to a few tribal longhouses bordered by one timber camp, far up the Baram River in a faraway land, in a fabled rainforest. Yet I could take you there in twenty hours from Seattle for a thousand dollars, and in many ways their problems are our problems. Alas, Borneo is next door. Or maybe, in the brave new oneness of this world, it is home.

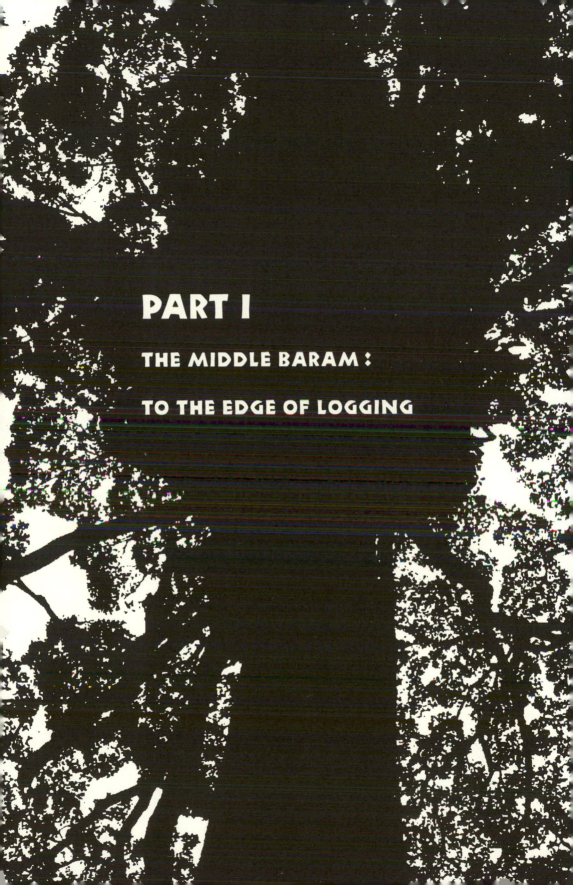

PART I

THE MIDDLE BARAM:

TO THE EDGE OF LOGGING

CHAPTER 1 🦎
INTO THE HEART

February 22, 1991

It's not easy, when you finally meet the enemy, and you like him.

It is night, giant cicadas and grasshoppers the size of cigars are banging against the screens and against the outside of the house. We open and close the door quickly, and mount the stairs to a large, white, empty living room on the second floor, above the offices of Tebanyi timber camp. Five hand-sized moths and butterflies are already inside, brown and white and black on the walls near the fluorescent lights. Tired from a day of bouncing in a pickup truck on rutted dirt roads, Fujino and I sink into the beige cushions of a chair and sofa. Newspapers are stacked on the coffee table: the *Sarawak Tribune* from the capital, Kuching; the *Asahi Shinbun,* from Tokyo. The room is a little bare, a little functional, made of planed boards painted white. It could be the lodge of a summer camp in Maine.

We are three days upriver by boat, or one long day by logging truck, all the way up to the edge of virgin forest, at the end of logging in Borneo. Fujino is the manager of the camp. He is fifty-five and looks thirty. Fit, wiry, smart, with an open smile. I have come with him from Miri, on the coast, in a Samling Timber Company pickup. "You're not from Greenpeace, are you?" he had said over the phone. I replied that I was a professor, open-minded I hoped, representing no organization, and that most of the publications on logging came from his opposition. I said I would like to hear the other side. What I did not say was that I had been in and out of his district for six months.

Now we sit, on the second night, in the headquarters of his timber camp. It is February, but in this rainforest a hundred miles north of the equator, all months are hot and wet. We have been talking of Japan. "The most closed country in the world," he says, laughing. "They should open their doors."

"To the rest of Asia?" I respond, incredulous.

"Yes. They have an obligation."

"But then," I protest, "it would not be Japan."

He smiles. In Borneo I have found a Japanese liberal, while to him I sound like a nationalist discovering the cult of the emperor. He puts water for tea on a hotplate. I realize I have rarely heard a Japanese say "they." In Japan it was usually "We Japanese feel . . . ," as if there is one heart, one mind. But he says "they." Fujino has been away a long time.

He came to Southeast Asia in 1964, he says, taking the pot off and pouring the water through a green-tea strainer into a ceramic cup. I am having Anchor beer, courtesy of Samling Timber, cold from the refrigerator next to the television. Fujino joined the timber import section of a trading firm right out of college in Osaka, and ever since he has followed Japan's tropical timber trade: the Philippines, Papua New Guinea, Indonesia, neighboring Sabah in north Borneo. Now most of those forests are gone.

"Are you learning Portuguese?" I ask, thinking of Brazil.

"No," he laughs. "I will retire."

Listening to the hailstorm of insects on the screens, walls, and roof after a day of rain, and smelling the sweet heavy cool of night, I am struck again by Fujino calling this "poor" forest. "Oh yes," he says. "This is the worst." That's why it's the last in Southeast Asia, I suppose to myself. "In Sarawak we get eight tons per acre average, less up this high. Sabah was fifteen tons per acre, Indonesia was more. The Philippines"—he seems nostalgic now, eyes distant—"went sixty to seventy tons per acre. You could practically clear-cut that forest, and it was better wood too. Mahogany, teak. Here even the eight tons is mediocre, half meranti and half mixed."

"Here you get eight tons per acre?"

"At best."

"But you say a good-sized meranti is five to eight tons?"

"Right."

"So you're harvesting only one tree per acre?"

"If it's a big tree."

"And that's all the marketable timber?"

"Right."

I look out the screen past the banging bugs into the night mist, thinking of what I've seen around his camp the last two days. The huge road cuts, the slides, the fifty bulldozers and front-end loaders scattered in the mud in front of the repair shed (over a hundred others are out in the forest), the trucks. And what I've heard and seen during the last six months in and out of the Baram: bulldozers appearing with no warning where people have hunted and gathered the forest for generations; Sarawak minister James Wong's position—"There are no native land rights"—repeated by timber companies to frightened natives in isolated longhouses upriver; traditional cultures forced suddenly into a new economy, damage to the forest. Now all this will come up here, to yet another virgin district, for only one tree per acre.

I know Fujino says they can make money on one tree per acre, but it seems stranger than clear-cutting. Why not leave the land alone? Does the world need the timber that much? This timber, the world's oldest rainforest? Should it come this cheap, just underselling Canadian fir? The natives' desires and rights do not enter the economics.

A single tractor team skids a log an hour from the forest, where the feller dropped it, out to the nearest logging road. There are fifty skidding teams working out of Tebanyi camp. A log an hour is an acre an hour, so the mechanical system of my world is advancing through the vegetable system of the natives' world at fifty acres an hour, here, at this one camp. To send logs to Japan to make plywood to build the Tokyo that we bombed—I can't help thinking. Suddenly I do not feel adequate to the subject, whatever it might be. I listen to a monster insect thumping its brains against the outside wall and open another Anchor beer.

It has started raining again. Within seconds the few drops have become a deafening roar on the zinc roof, making conversation impossible. We wait a few minutes; it subsides.

"The natives don't want this to happen," I say absent-mindedly, looking at the window, lost in the screen's kinetic art: white bug bellies crawling on a field of black damp.

"Don't want what?"

"The logging. Even the ones who work in your camp. They won't tell you, but they tell me. They wish the logging had never come."

"They agreed to it. All the middle Baram longhouses." I look around, surprised at my friend. Does he believe that? He is staring into his tea.

The middle Baram has seven longhouses, each with an elected "headman" or chief, and one district chief called a Penghulu, from the main longhouse, Long San. I have been told often over the past six months how the headmen and the Penghulu have been given the authority by the government to sign legal documents on behalf of the longhouse. Now the headmen are bribed by the timber companies to sign. I know what agonizing divisions this is causing in the longhouses. An elected headman can be removed, but they are aristocrats, and upriver, traditional authority is much respected; opposition to authority is antitribal. After all, perhaps two hundred people must share one porch, deep in a jungle. Criticism of the headman, much less removal, is a trauma for the entire longhouse. Opposition to logging, then, often means opposition to a headman's decision and so runs counter to *adat,* or customary native law, and carries a heavy stigma. Little wonder that longhouses are under great stress.

All this is going through my mind, and more: in November I had read the middle Baram timber agreement with Samling, while angry natives surrounded me, pointing at their own names, saying they never signed. I am very curious to hear how Fujino handles this subject, especially since his company says "There are no native land rights," and then goes to great lengths to obtain logging agreements on this same nonnative land. I take two sips of beer to calm the mind and hold the tongue, a habit learned in Japan. Finally I look at a butterfly on the wall, big as Madame Butterfly's foot, and ask: "The headmen *and* the longhouses agreed to the logging, or just the headmen?"

Fujino studies a banana in the dish. He picks it up. "The agreements are signed by every head of household—there's a list of names attached to the agreement; you can see it."

I like Fujino a lot. He has been kind and open, and I think a good deal more of Samling Timber Company for having met him. But either my information is wrong or he does not know what his company is doing, or he has no idea how much I know and for the first time he is lying to me.

"I was told at Long San," I say, "that the middle Baram longhouses, every one of them, rejected the agreement in 1986, in '87, and again in '88. You're proceeding under the '88 agreement. Only the Penghulu, the headman, and the two committee members signed. And I was told that they were paid. By Samling Timber."

Suddenly his eyes grow dim. He peels the banana, says nothing, and will not look at me.

"That list of names attached to the agreement," I continue, "are you certain that they had read the timber agreement and were signing? Or is that just a census of the longhouse attached to the contract against their will?"

His eyes are misty. No reply. He studies the banana, then looks away, and we sit a moment in embarrassed silence. He knows he has been caught in a lie, and he is ashamed. God bless him. In all of Sarawak and Tokyo, I will not meet another timber man who shows shame, and I realize instantly that if I ever write this story I will have to use Fujino, my friend, to get to its heart. But then, we are not meeting because of the justice in this world.

I flip through my notes again, wanting to change the subject. "If you had a chance to reply to your opposition, to native activists and the environmentalists, what would you say?" He hesitates. We drink some beer. The rain has stopped, the eaves are dripping; one cicada honks. He speaks in oddly disjointed sentences:

"Land rights are an internal problem for the Sarawak government. My job is to execute what the boss directs. You have to remember that a free market and overproduction push down the price. We're lucky if we get half the possible timber out of an area, so we're not making so much money. You can't calculate profits at 100 percent production at top price."

"How much time is left here?"

"The best is already gone. In ten years it will be finished."

The fact that as a reply to the opposition all this has missed the point is point enough. I don't want to badger him. We drink tea and chat about fishing in Montana, my homesick memories filling his fantasies of living again up north: cold streams, bright trout, crisp air with the steel smell of snow.

We are about to go to bed. I hand him a one-page summary of the history of native land rights, which I had written in November to show to ministers and lawyers on both sides of the issues, to measure the differences in perception. To my surprise, he is interested, and wants to look at it right away. He reads with obvious, painstaking care: since 1958, "tribal land can pass into individual ownership (and out of the tribe); existing customary rights can be extinguished by the government ('six weeks from notification in the *Gazette*'); timber concessions on tribal lands are granted in Kuching with no prior consultation or

Land law in Malaysia is a state prerogative. There is widespread disagreement on current law, from James Wong, Sarawak Minister ("There are no native land rights") to Thayalan, CAP lawyer ("The constitution says native customary use establishes ownership; therefore all timber licenses on customary lands are illegal"). Many see the law itself as subordinate to political will in Sarawak. Certainly the political will of the last one hundred fifty years has been to diminish native control and enhance state control of land in Sarawak. No recent court case defines native land rights.

History: Under Brooke rule (1841–1946) the European system of private and state ownership of land was superimposed on top of native customary law (*adat*), and "the evils of shifting cultivation" were to be "eliminated" (Circular 12, 1939). In practice, however, the Brookes discouraged development and encouraged native autonomy. Charles Brooke (1915) warned natives that they should keep their land rights at all costs. The seeds of legal appropriation by the state had been sown by 1939, but had borne little fruit.

Traditional native land use to this day recognizes four land divisions: (1) The outer boundary of *tribal land* extending to the ridge top on both sides of the river and to the boundaries of the next longhouse; therefore, along the river, all land is claimed by tribes (the interior is usually Penan ancestral land). (2) Within this boundary is a *forest preserve,* open only to hunting and gathering or the occasional communal felling of a tree. (3) *Usable forest,* where trees as well as other forest products can be taken by individuals, and new land cleared. (4) *Customary land,* which has been cleared for cultivation and belongs to the original feller and his descendants as long as it is worked (including fallow periods during crop rotation).

Since 1952, the land code has been changed in many ways: new customary land can be established after 1958 only with permission from a district officer; traditional native land claims are not de facto, but must be requested from and granted by the state; tribal land can pass into individual ownership (and out of the tribe); existing customary rights can be extinguished by the government ("six weeks from notification in the *Gazette*"); timber concessions on tribal lands are granted in Kuching with no prior consultation or even notification of affected tribes.

In support of native land rights: customary rights remain recognized in sections 5, 10, 15, and 18 of the constitution; timber company agreements and legal briefs recognize tribal land boundaries; forest preserves may be protected by law, though in practice not a single request has been granted.

Summary: Natives have lost almost all control over the land they and their ancestors have inhabited for generations.

even notification of affected tribes." As he reads, I busy myself picking up the room, looking at the Japanese magazines, *Jump and Friday,* on top of the VCR. I know the last sentence of my page, and out of the corner of my eye wait for him to reach it: "Natives have lost almost all control over the land they and their ancestors have inhabited for generations."

Finally he finishes, puts the page down and looks up. "I have never heard this," he says slowly and — I believe — honestly. "That's very difficult for them."

He has been in Southeast Asia for thirty years.

Later, in bed, alone in the three-bed dormitory room of the timber camp, I turn on the light — Samling's generator runs full time (for refrigerators, computers, yard lights) — open my journal and listen to a hundred unknown noises in the night. No rain. I am almost finished in Sarawak, on my way back home to fishing in Montana. My mind drifts back over journeys first with family, then with my wife, then alone, and I recall especially the series of events that, four months before, in the fall, had led up to the meeting at Long Moh. That was when the native activists and Samling Timber met head on. That was when, at the top of the upper Baram, Samling asked the natives to sign another timber agreement. The *middle* Baram had already signed, and logging had started. But Samling must look ahead, and the *upper* Baram, almost the size of Massachusetts, was next, and one of the few large concessions left in Sarawak. Samling needed a timber agreement with the upper Baram, although "there are no native land rights," and the upper Baram, with a few extra years to listen and learn, had proved a good deal more ready than the natives downstream had been.

That meeting, November 27, 1990, was when it was visible, when the gap between our world and theirs was bridged in a moment of contact clear as an arc welder's torch — if we can look at it. That was when Mr. Sei of Tokyo, regional manager of Samling Timber, sat down on the Long Moh veranda with the upper Baram Penghulu, seven headmen, and a hundred natives. At that time and place, the insatiable appetite of Mr. Sei's people and my own for the resources of this earth met some of the last hunting-gathering-farming inhabitants of the Southeast Asian rainforest, a going concern for over a hundred million years, with about ten years left. For a few hours on that porch the children stopped flut-flutting up and down bare boards on bare feet, the women stopped weaving rattan mats, and the twentieth century happened. The dogs knew something was up and stayed clear. Only the roosters, strutting in the security of their little pea brains, were safe from the sense of change.

Consider the gap bridged on that porch. The ten largest banks in the world are now Japanese, fueled by the Tokyo real estate market. The dollar value of the land in Tokyo, at 1990 market prices per square meter, was equal to all the land in all of the United States plus all the stocks on the U.S. stock exchanges. The capital generated from those paper assets is funding much of the world (including America's debt service), as well as Japan. The tropical timber trade is pursued by a number of Japan's biggest import firms, each associated with a major bank. Those trading house conglomerates—Marubeni, C. Itoh, Nissho Iwai, Sumitomo, Mitsui, Mitsubishi—in order of their 1990 imports from Sarawak, come to a poor nation in Southeast Asia. They have money to offer for logs. They drive very hard bargains, but they can pay for any volume the country can supply. If the country cannot guarantee a large and steady supply of logs, Japan will shop elsewhere. The Sarawak government "gives" a timber concession to a politically connected person; that person retains a local Chinese timber company to do the logging; one of the world's largest banks, in Tokyo, says to the Chinese company that capital investment is no problem. Need three hundred bulldozers? Pay us back in logs. Meanwhile, it has been up to the state government to clear away any impediments such as native land rights, which have disappeared in Sarawak since the coming of logging to Southeast Asia. So when Mr. Sei of Tokyo sits down on the porch, representing a Sarawak Chinese company tied to Hong Kong funding (including Citibank) and Japanese import firms and banks, he has a leverage it is hard to imagine. Those firms would not have come to Sarawak if the ducks were not lined up.

On the other side of the porch, Mr. Sei faces seven headmen and a Penghulu. They all live upriver, in the longhouses. As aristocrats, they may control a good deal of land, but they are still well off only within the terms of the local economy. In that economy, in upriver Borneo, rubber is the main and often the only cash crop. An acre of tapped rubber trees around Long San earns about $2 a day; a family tapping all it can manage might earn $170 a month. Into this world comes Mr. Sei.

I was told that Samling promised each of the seven headmen of the middle Baram $200 (500 Malaysian ringgit) a month for the life of the concession, to sign. And Samling gave them $200 cash up front. This was all hush-hush, of course, and later the longhouses refused to ratify the agreement. But the headmen had already signed. To the headmen, this was big money; to Samling, it was not a significant investment for a district that would yield $3.6 million of timber, gross, per month.

The idea of a Japanese import house associated with one of the world's largest banks, a Chinese timber company with its Hong Kong backers, and the Sarawak government, sitting down with seven natives who have been told they have no rights, and are not allowed to have counsel present, seems unfair.

I lie in bed in Fujino's timber camp, looking back on what I've seen, and turning old pages of my notes. September, October, and November seem to lead to Long Moh, where the natives sat face to face with Samling Timber, and with their own headmen. It was a beautiful longhouse, near the very top of the Baram, with flowering shrubs and an old, six-foot drum hanging from the roof of the porch. They beat the drum to call the meeting together.

As I wait for sleep, I remember one night in particular on the way up to Long Moh, because it was the first time in weeks that I had taken a few hours to slow down and think, and because my thoughts had turned mainly to our young guide and interpreter, Richard. We had stopped at Long Anap, halfway up the Baram. We were heading up to Long Moh in a sixty-foot longboat, picking up opposition leaders like hired guns. "The Magnificent Seven," I called our expedition, and Richard countered immediately, "More like Mission Impossible." I stared at him. This was a boy right out of the jungle and the longhouse. "I've seen reruns," he explained. His English was excellent — from the schools of this once-British colony. While our leader Joseph Wang Tingang took charge of the gathering in the longhouse room, Richard translated for me. I flip back and find the page in my journal.

It was ten o'clock on the night of November 25, at Long Anap, Richard's home. We sat cross-legged under a kerosene lantern on rattan mats and linoleum. In the room were: four children against the wall; five older men and four older women on the floor (these in shorts, slacks, sarongs, polo shirts, tattoos, women with earlobes stretched by brass rings past their shoulders and arms tattooed solid blue from knuckles to elbows, some men with traditional bowl haircuts and short pigtails); and on the raised dais, another six men as well as strange bearded me, sole object of interest to the children. From this dais, Joseph led the meeting for over an hour, in Kenyah. The discussion became heated, never with shouting or interruption, but with quickening, articulate intensity.

"Will you win at Long Moh?" I whispered to Richard. "Will they turn down the logging agreement?"

"I don't know. Joseph says the longhouses are strong, but the company has been going up there for weeks. It's hard to say what the headmen will do. It will be a fight."

"If the logging is stopped," I asked, "what will the longhouses do? Live the old way?"

He smiled and shrugged. "No. Live their way." He leaned back and rolled his clothes up into a pillow.

I thought then of the young, those strong, handsome Borneo teenagers, men and women looking at times Malay, at times Mongolian or American Indian, at times white. What will it mean, after the logging, to be a native in Sarawak? An Iban, Kayan, Kenyah, Penan?

Richard was soon snoring away on the floor. Swept constantly and wiped with a damp cloth after meals, no shoes allowed, no furniture, the thirty- by forty-foot hardwood floor, opening up to rafters and a high roof, was empty—away from the gathering—and clean. In that heat, with no need of blankets, bed is where you lie down, and as the meeting droned on, one by one people voted with their backs.

Beside me, Richard was gone for the night. He was twenty-one years old, a handsome, smart boy. His mother was Penan, from Long Bangan on the Tutoh River, his father Kenyan from Long Anap. Richard had moved in with his mother's relatives in 1988 as the logging came through, and he had worked in a timber camp. But by November the logging had finished in that district, and he returned with my wife and me from the Tutoh to the Baram.

His Penan relatives on the Tutoh, including the activist leader Jewin Lihan, had vigorously opposed logging for years. Richard's young life on the Tutoh and Baram rivers was already inextricably intertwined with both logging jobs and logging opposition, and in the last two weeks of November he had pretty much seen it all. He had helped his cousin as a substitute sawyer; we went into the forest and watched him bring down a large meranti.

Lying in the timber camp, almost home, I think of Fujino's words: "That's very difficult for them." I think of the young and of those weeks in November with Richard. Richard, I note in my journal, the next generation—that is where I should begin the story that leads to Long Moh.

CHAPTER 2 🦎
GOING TO TOWN

November 11, 1990

In the darkness, the first cicada begins far away, a buzz in the midst of sleep. It comes closer, whirring and rasping, and others take up the call, bugles in the morning silence. Beneath the longhouse of Long Bangan, one rooster challenges and half a dozen respond. Somewhere among the wooden pillars, a baby pig squeals. An old dog springs awake; his nails click on the wood as he races stiff legged over the sleepers and out the door, touching once on the ladder, five feet to the ground. More dogs begin to bark.

Richard opens his eyes. Beside him on the wooden pallet his nine-year-old niece and three-year-old nephew are curled against the morning chill, the single wool blanket kicked to the foot of the boards. Slits of gray light run the length of the wall between the planks, as far as one can see in the gloom: fifty feet, a hundred feet, then two hundred feet. Minute by minute the jungle dawn lightens the damp mist sifting through the longhouse. Hanging baskets and hoes appear, a fifty-gallon oil drum, a Panasonic tape player, knives, blowpipes, and guns, and as we lie watching, sleepers take shape. Most of Richard's family—over twenty members—are in sight, some on raised pallets, many on the floor. A wisp of smoke rising from the alcove to our left shows that his aunt has started the fire.

His feet touch the floor without a sound. In Adidas tennis shorts and nothing else he pinches out the wick set in diesel oil that has kept mosquitoes away, and creeps toward the kitchen. Breakfast is early for him this morning; he is going on a trip with us. It will be a long pendulum swing down the Tutoh, then up the big river, the Baram. At the bottom of the pendulum we will spend a few days in the town of Marudi, then journey two or three days up to Long Anap in the middle Baram, Richard's home. Long Anap is scheduled to be logged within the year; it is a three-hour walk down from the ridge where Fujino lives in Tebanyi timber camp.

It is lucky for us and for Richard that we can travel together. We want to go up the Baram—beyond transportation. This means finding your own canoes and outboards and boys to run them (the equipment is not rented to strangers). Richard can be our leader, guide, and interpreter. At twelve dollars a day, it is one of the best jobs upriver.

Richard is excited, though he does not show it. A visit to Marudi, sole town and center of the Baram universe, is a big deal. So is going home. We are excited too; after weeks in logged-over areas on the Tutoh—all virgin eight years ago—we have found a guide to take us up the Baram, beyond logging.

But we have no idea at the time how far up we will finally go, or how deeply we will become involved with native opposition to the logging.

We are not moving quickly in the predawn light. The night before last there was a farewell party a few miles away at Jewin Lihan's new camp, ending Richard's two years with his relatives, and we still have not recovered. In the midst of six or eight bamboo huts on stilts, hidden in the forest, there was traditional solo dancing to a tape of lute-like *sape* music, slow twisting dances of great grace and dignity, led first by Jewin's father, the tribal elder, then by the present headman Jewin, and then followed by everyone in turn, including women and children. My wife, Juliette Crump, known to Jewin as a professor of dance from America, was given the honor of following the headman. I thought it interesting that they placed her ahead of me, and thought also that it never would have happened in Japan. The hornbill feather hat was fitted on her head and she stepped slowly to the center of the dirt circle, lit by a fire and a kerosene lantern. She had just seen the dance twice, and she gave a spirited imitation and interpretation. She earned long applause; Richard was impressed.

Traditional dances were followed by several hours of serious bop: "Jailhouse Rock," "Hound Dog," "Heartbreak Hotel"—the entire Elvis collection on tape—their tapes, and their Panasonic player. Those few of us in the clearing who had gone to North Carolina for college in 1959 were taken back to some ancient customs of our own. Niece Linda Bevis taught Richard new steps and daughter Sarah Bevis led processions of children weaving in and out in the style of line dances around the world. Some of us—the older ones, I think—began to fade. We climbed the five-foot ladder up to the sleeping platform and wriggled into sheet envelopes on the springy, open bamboo slats of the floor. The Beach Boys had just begun to sing. We lit Shell Oil mosquito coils. Richard and other serious party animals carried the boombox out into the jungle, away from camp, where "Surfin' Safari" blended easily with a thousand other noises of the night.

And always the children. On our little ten- by fifteen-foot floor, between the open slat walls and a zinc roof, the children encircled us, a dozen of them, from the moment we climbed the ladder until we were fast asleep. Some spread alone to corners, some bunched two or three deep, watching us silently out of big eyes and beautiful faces, then erupting all at once in laughter or pokings or exclamations at something we possessed or had done, some gesture, some word, some artifact slid from a pack as we ate, brushed teeth, found a headlamp, undressed, opened white cotton sheets, rolled pants into a pillow. As we closed our eyes, I could feel their stillness. I did not hear them leave.

After the party—yesterday—Jewin's pet myna bird awakened us at dawn. We ate our breakfast of rice and gathered for good-byes. Jewin's father was headman before him; a beautiful, diminutive old man with traditional tattoos and stretched ears, he had, without a word of English, been with us throughout our visit. As tribal elder, he had led off the dancing, and he and I had spent hours sitting on the edge of the hut platform plunking gourds with his blowpipe, children excitedly retrieving the darts. Among other presents, we had brought him a large map of the world, and when we left, he hugged each one of us, pointing very slowly and repeatedly in each of the four directions. Then he made a gathering motion into his stomach and spoke while Jewin translated, "Go, go far, in every direction, and then come back, come back." He seemed, without maps, to be telling us about the world—that it is very large, and very small.

We hiked out through the forest, back to the clearing at Long Bangan. Now, this morning, a few relatives are rising to send Richard and us downstream to Marudi. His aunt has built a fire in "the kitchen," a covered porch extending out from the back side of the longhouse. Her extended family, like most, has built its own kitchen adjacent to its living quarters in the longhouse. The cooking fire is laid on the floor in a contained bed of ashes a foot thick; pots and supplies are slung from wires above. A jar of water, a few shelves with utensils. Most of the smoke, which neither stings nor smells, goes out easily through the slat walls and eaves; the rest keeps mosquitoes away. We sit on the floor to a breakfast of weak coffee heavily sugared, white rice on a banana leaf, dark meat chunks that come from "some kind of cat," and sago: a gelatinous gray mass of pure starch pounded from the trunk of the sago palm. "The food is not what you're used to," Richard says, smiling. No one else around us understands English, and we all laugh. "What will you do at Long Anap?" I ask. "I don't know," he answers. "Maybe work for Samling."

Richard has washed and slicked back his black hair. With his high cheekbones, fair complexion, and flashing eyes, with the beautifully clear, hairless skin typical of Borneo natives and much of Asia, with the smooth muscles of his bare chest and arms, he really is a hunk, Linda mumbles through her sago. He also has a quick, intelligent, and ironic smile. A century ago I would have wondered if I would get all my womenfolk out of this jungle. Now, Beach Boys echoing in my head in the midst of thatched huts — "Barefoot and Bad in Borneo" was Sarah's title for the party evening — I really don't know what to wonder, but I do.

After breakfast Richard straps on a parang (the upriver machete carried everywhere), grabs a T-shirt and the little Naugahyde satchel that contains all his clothes and belongings, says quick and casual good-byes and climbs down the ladder. Lucas Paran, our kind host, waves good-bye. We shoulder blue and green canvas backpacks and fall into line, the shoeless and bare-chested Richard with his satchel, brunette Sarah and blonde Linda in shorts — to hell with leeches — the white-haired, blue-eyed Juliette looking tough in green Safari pants and a Beijing Dance Institute T-shirt, and old bearded me in khakis. We walk out of the clearing into second-growth forest.

When the big trees were taken out three years ago, leafy vines (like kudzu) overran the scraggly, half-cut forest, covering and strangling half the trees. They don't clear-cut in hill country. The trees are so various, so many species are unmarketable or small, that they pull out only prime trunks. The numerous connecting vines catch on the falling trees and on the bulldozers dragging them out. Then it rains, and in months, weeks, in the unprecedented sunlight where once the giants stood, entirely different kinds of vines and ground plants choke the forest.

Long Bangan was created by the government; these are resettled

Penan, who ten years ago were semi-settled and depended on hunting and gathering. Now the logging has come through, and while the government has helped build them a longhouse (in a clearing, although they hate the sun), they have no economy; their staple meat was wild boar, but they have not seen one for three years. The monkeys and other lesser animals, which they eat, are gone; sago palms are very hard to find. Primary jungle is three days hard walk away, and since game flees from the sounds of bulldozers and chainsaws, good hunting is much farther than that and retreating almost a mile a month. No one bothers to go anymore; you could not carry enough meat back. The second growth is rich with leaves but supports surprisingly few animals, which thrive instead on the nuts and fruits of primary forest. I ask about the deer, which love our American clear-cuts; some species can browse second growth, most do not. Even if the deer came back, second growth is almost unhuntable, says Richard, because it is so thick that game cannot be sighted or approached.

As we walk through trees draped with strangling vines, the cicadas make conversation difficult—some honking like cars, some shrieking like banshees, some whining like a spoiled child. We come to a long, skinny, slippery log bridging a stream. Richard starts across without thinking, then realizing he's alone, stops, nearly to the other bank, and turns around, still on the log. Mind you, I have hiked and climbed a fair amount. I have experience. I have Vibram soles. But a tree skinny as a telephone pole laid bank to bank eight feet above a stream? With no bark, polished smooth by bare feet, in a light rain? I look down— the stream is too deep to wade. I look up—no hanging vines, nothing to hold on to. I even check for a Tarzan special, and for a moment can hear the Penan laughing around the fire as one recalls the huge, awkward graybeard who swung across with a blood-curdling yell . . . Backpacks would make sliding across the log quite tricky. Not to mention pride. I softly curse Richard's splayed bare feet, toes curled to the log. He smiles, turns, and continues to the other side, and waits without a word. He knows perfectly well that language is part of neither the problem nor the solution. I know it too. Then, to my utter dismay, daughter Sarah walks across. Then Linda, who took thirteen years of ballet, and Juliette, the goddamn professor of goddamn dance. They just walk across. I curse silently, since in Asia cursing is seen as a weakness. It is. I am. Concentrating on not cursing, however, distracts me enough to begin, two steps, three . . . There's no going back. At the thought I stop, of course, and begin to waver, then three more steps, then the middle of the log above the middle of the stream and panic, knees bent, arms flailing, leaning back and forth like a high-wire artist. The thought that total confidence is the only cure panics me even more until I break into a little sideways shuffling run which looks like the Three Stooges doing ice ballet but lands me in many arms on the opposite bank. Richard is laughing openly. That seems permitted in Asia, at least among the Penan.

We slog through a wet, overgrown path, walk the length of a huge trunk (with bark) fallen across a swamp, and ascend the muddy side of a ravine to the road. It occurs to me: that was the main entrance to Long Bangan. I have been living with three hundred people who take this for granted. Well, I'd like to get them on skis.

On the dirt logging road, Jewin Lihan waits, furling his black umbrella as the sun comes out. He speaks to Richard in Penan. "Here are the champion dancers," Richard translates. We thank Jewin again for the hospitality at his camp, and ask about his father's health. Jewin is head of the Association of Penan, the tribe closest to the jungle and to hunting, the tribe most affected by logging, and the tribe that began the blockades of logging roads. He and some followers split off from Long Bangan in April to build a new settlement, where we had our party, closer to a school (one and a half hours' walk). Jewin looks tall and handsome in bare feet, black slacks, white shirt, and hornbill feathered straw hat.

He stands on the empty dirt logging road carrying shoes and a small satchel of town clothes, for he also is going on a trip. His father must have gestured hard to the north. Jewin is heading to Japan, where he will testify against logging at the meeting of the International Tropical Timber Organization in Yokohama.

We hail a Baya Timber Company truck driven by local boys, who gladly pick up their friend Richard and give the leader of logging opposition in the Tutoh his first lift toward Yokohama. After half an hour of bouncing in the mud we reach Long Tarawan, a huge new longhouse with a modern district school, and descend a notched log to the dock on the river. Skies have cleared. Jewin finds friends who have room for one person in a dugout; they are going down to meet the express boat. He puts his satchel and shoes aboard and climbs in.

A boy pulls the rope on a Suzuki outboard, and they roar off. We will meet up later, when we catch a ride.

Absolutely nothing is happening at ten o'clock in the morning on the dock at Long Tarawan. The brown Tutoh slides from here to there, a hundred yards wide; now and then an outboard races by, rocking the four dugout canoes tied to the dock. The dock is planks nailed to two huge logs; the logs are held by cables to the bank.

We receive a visit from the retired schoolmaster at Tarawan. A small, dark man with white hair, barefoot, he carefully descends the log to the dock. He has seen the foreigners and has come to practice his excellent English. He remembers the third Rajah, Vyner Brooke, and the Japanese occupation during World War II. No, he says, the Japanese did not make it up this far.

He leaves, and we have a choice of being eaten by the biting sand fleas on the dock, or of eating cookies in the sweltering innards of the Chinese storeboat moored beside us. It is the only store within three hours by river, or two days of walking. The boat is thirty feet long, with a wooden cabin running almost the entire length, painted a peeling green. On the tarred roof, eight drums of oil and gasoline are lashed. We walk the short plank from the dock in through the cabin door, right behind the front window, squeeze by the captain's wheel and levers, and are in the store. There are no other windows, although a door opening to the stern at the other end gives some light and ventilation. Still, a small cabin with tarred roof, on the equator in the middle of a sunny day—I am sopping wet. A narrow aisle runs the length of the cabin in between shelves of canned goods, cookies, crackers, sacks of rice and sago flour, condiments, fuel, mosquito coils. If I stand up straight my head hits the ceiling or my nose hits the kerosene lantern swinging in the middle of the corridor. From a "cooler" with river water we fish out tepid Cokes, and take a package of cookies off the shelf. Aft of the goods at the end of the aisle is a small desk, where sales are recorded in an old ledger book, and behind that is a four- by eight-foot space with a trunk and a hammock, where the owner/operator lives.

He was living there when we entered and is living there now, lying on his side in the hammock in the stifling heat, his practiced eye able to tell we have finished our cokes and cookies and are bored, but not buying anything else. Time to pay. He sits up on the edge of the hammock. He looks about forty, and speaks some English. He inherited the boat from his father, who was born in Miri on the coast, his great-grandfather having come over from southern China in the mid-nineteenth century. Did his great-grandfather participate in the Chinese rebellion against the first Rajah Brooke in 1857? No idea, but he knows quite a bit about the incident. His ancestors first settled near Kuching. The boat, which his father built, has been on the Baram and Tutoh rivers for fifty years. His father couldn't navigate the fast water above Marudi, but the son put in a new engine fifteen years ago and this was the first storeboat to ascend the Tutoh. He remembers the crowd of children sliding down the bank here at Tarawan the first time he docked. He returns to Marudi for supplies as little as possible, ordering most goods up on the express boat to the end of its run (three hours down from here, but a day and a half back up). Once he goes downstream, he must come up very slowly, working the eddies along the bank and sometimes waiting out rains and high water for weeks at a time. Getting all these goods up to Tarawan, which is as high as he goes, is a big investment. He has been here three weeks. A good Confucian, he knows every item in his inventory, and I can tell that his books are in order. The Chinese dominate the economy of Malaysia; every logging firm in Sarawak is Chinese. He lies down again in his hammock and closes his eyes. But I know he can hear his shelves.

Richard returns; he has been off scouting a ride downriver to the last express stop. Still three hours above the scheduled express service, we are completely on our own. No roads, no telephone, no public transportation. Richard has returned from the longhouse with a fourteen year old carrying a fifteen horsepower Yamaha outboard on his shoulder. The two wrestle it to the back of a thirty-foot hollowed-out log and plug in the gas can. Climb in, Richard motions. He takes the front, I and the three ladies are in the middle, the boy at the rear pulls the starter rope and we are off downstream. If you put a hand on the gunnel, two fingers are in the water. In other words, we are three inches from being a sunken log with a motor. A Borneo submarine. "A full log?" I ask Richard. "A full log," he replies. "No problem" comes trilling from the pipsqueak in the rear who is stealing his father's log to make a thirty dollar profit off the windfall that has downed my family on his dock.

"Does the Tutoh have crocodiles?" asks Linda the lawyer, ever mindful of fascinating possibilities that ordinary people might overlook;

Richard turns around to answer, as he usually does when the young ladies speak, and the right gunnel dips in a few gallons of water; "No problem" sounds immediately from the eighth-grade captain as he counterrocks the canoe and guns the motor. "Yes," says Richard, and we are certainly glad that matter is settled. We watch the brown water slide by, wetting our elbows. "Chomp, chomp." "Sarah, stop that please."

In two hours we reach Long Panai, where the afternoon express boat has turned around, waiting to run back down to Marudi. Jewin Lihan is waiting on the dock with friends; we join them. Now, for the first time in weeks, we can actually buy a ticket and go somewhere — Marudi at least — on a schedule. Vaguely, deeply, I feel that we are back in civilization; and so, it occurs to me, I must think civilization is where events can be predicted by means of a watch. That, it further occurs, might severely limit the kind of events my civilization keeps track of; might define, might even create the events my culture values. Goodness. My watch is buried in my pack; I cannot remember which pocket.

The dock shakes; the express boat roars to life.

All over Malaysia, up rivers and for short ocean crossings, the express boats are the same design. For those expecting the African Queen, it is something of a shock. A sleek iron bullet a hundred and ten feet long and twelve feet wide, the silver boat looks like the body of an airplane stripped of wings and tail and set in the water; on top, add a few planks and four-inch railings for lashing boxes, spare propellers, chickens, and tourists to the roof. The boat looks lean and mean, and when the giant diesel revs, you know it is. The entire body vibrates as it backs out from the dock, then a deafening roar as the mud churns up, the wind begins to whistle and we're off downstream, wake lapping the fern-lined banks far behind us.

We crazy whites sit on the roof with baggage and chickens to watch the cut jungle roar by at 30 mph. Richard, Jewin, and every other native take cushioned seats in the air-conditioned, airplane style cabin and watch the big Scotsman in kilts beat up a woman in a black gown slit to her thigh — professional wrestling on videotape. The TV at the front of the cabin is decked with flowers, as if it were an altar.

There are only five or six towns and two big river systems in Sarawak. The river in the south is the Rajang; near it are two towns, Kuching, the capital, and Sibu, the timber port for the Rajang. The river in the north is the Baram. Near the mouth of the Baram is Miri, an offshore oil port since 1910 (Shell Oil), and now the timber port for the Baram. A fourth of the way up the Baram is Marudi, the only town upstream, population three thousand, just below the junction of the Baram and the Tutoh.

The Tutoh is the main tributary of the Baram; between them the two rivers drain the entire northern highlands. By 1985, all the lowland rainforest in Sarawak was gone. In the 1980s the logging went up the Tutoh, home of most of the nomadic Penan; about 85 percent of the Tutoh drainage is now concessioned, about 75 percent roaded, about 50 percent actually logged. With all this activity, the sleepy colonial outpost of Marudi has boomed.

To folks upriver, Marudi is the big apple, and Richard is happy to be there for the third time in his life. Five blocks of wide, dusty streets and Chinese-owned general merchandise stores and restaurants sprawl back from the wharf, where this morning thirteen express boats are tied up, most from downriver. There are cars, three times more than last year, and motorbikes, all brought up by barge, for no road goes out to another town. People drive twenty minutes in any of three directions, and then come back. Everything you could want spills from the stores into arcade displays, sheltered by arches and awnings from the blinding equatorial sun: flour, cornmeal, sago flour, Tupperware, flashlights, sarongs, T-shirts, tennis shoes (Adidas and LA Gear) Stayfree napkins, M&M's, badminton rackets, swim masks! (the old methods of river fishing included diving with spears), cameras gathering dust, Levi, Lacoste, and Benetton imitations from Thailand (wash separately), bins of open food: gingerroot, dried squid, garlic, onions, dried fish the size of a fingernail; Magic Dentist paste (don't ask—we watched a curbside dentist pulling teeth with pliers), guitars, Lux soap, Guinness malt, beach balls, plastic M 16's with camouflage jungle scenes on the cardboard backing, heavy-duty wrenches and pliers and picks, outboard motors and chainsaws—Husquvarna and Stihl. And for general interest in the street: two dogs stuck together and loudspeakers, strung all over town, playing "Will You Still Love Me Tomorrow?"

It is a thrill for us, for we are back in Fort Benton on the upper Missouri River during the fur trade of the 1840s or the cattle boom of the 1880s. Boats whistle and come and go, goods are unloaded, old men from upriver walk down the street, some with feathers and tattoos and long ear pendants and loincloths. I remember George Catlin's wonderful paintings of the Indians at the Missouri River forts in 1832, and L. A. Huffman's photographs of Miles City, Montana, in 1884. Catlin thought the natives were beautiful and dignified; that their land and economy were being wrongfully taken, that huge tracts of the West should be set aside permanently for the Indians to enjoy in their traditional life or to develop as they wished. A pang strikes my heart; a tug is going by in midriver, pulling a mile-long raft of logs. Their wet brown backs glint in the sun. The buffalo are being slaughtered again.

On the wharf, Richard acts cool among the hustling Chinese shopkeepers, the crates and cartons. Within minutes he has encountered two friends, and we can tell they are making heavy plans. We can also tell that he enjoys being found in the company of Linda and Sarah. When he huddles with his friends, however, there are no knowing glances or jokes. The natives are very polite about such things, as are most Asians. The women say they have felt no threats, no direct pressure, no hassling west of Seattle. As we talk, we are having a cup of coffee at the cafe facing the dock, watching Richard get his town legs.

Our first business in Marudi is to find the office of Friends of the Earth, Sahabat Alam Malaysia (SAM). Jewin has already gone there. SAM is sort of legal, and sort of not legal. Like the communist party in America, or the Weathermen, or the Black Panthers, they exist, and they are watched, hassled, and sometimes jailed. We don't want to ask strangers directions; we may well be observed doing business there — then visas will be checked, travel permits pulled. In Marudi we must remember that storeowners are Chinese, and that most Chinese support economic development, including logging. We are out of native territory.

I have been told where SAM is; Richard does not know. Off the main street, in the rear of a row of shops, with no sign at all, a little blue door leads to a flight of narrow, dark stairs; at the top, on the left, is a plain door. It creaks open into a small office. If Bogart and Bacall were in Borneo, surely they would be here. A secretary is typing on an old Royal; five natives in loincloths and feathers are waiting to talk to Harrison Ngau or Thomas Jalong or Raymond Albin. A lovely Kenyah woman, Nancy from way up at Lio Matoh, rises and comes forward to lead us in. Jewin is already there; now he has his shoes on.

SAM is paying Jewin Lihan's way to Japan, along with Thomas Jalong and three other Sarawak natives. He has old friends here with whom he has worked for years. We walk through the tiny reception room (no rug, no air-conditioning), to the second partition that creates an office twelve feet square, one bookcase, one phone, a fan, one window with a view of . . . the same two dogs, relocated but still engaged. Richard is excited; though he has been to Marudi twice before, he has never been to this office, and has never met Harrison Ngau.

Harrison stands behind his desk and extends a hand. He is about five feet seven inches tall, young, chunky, and strong, with an engaging smile. He wears the open white shirt and dark slacks of the tropics.

Harrison Ngau is one of the most effective activists on earth. *Time* magazine, in its Earth Day issue of 1990, recognized him after he had won a Goldman award in San Francisco as a "grass roots hero" for saving the environment.

Harrison Ngau, a Kayan tribesman in Malaysian Borneo, has endured imprisonment, house arrest and government harassment over the past three years. His "crime": helping Borneo's indigenous people try to halt the rampant logging that is destroying their way of life and some of the earth's most ancient tropical forests.

When timber interests first came to Ngau's area in the state of Sarawak in 1977, several thousand natives lived entirely off the forests. But logging and settlement plans have reduced that number to fewer than 500 Penan tribesmen, who still cling to nomadic ways. Even these remaining nomadic clans are threatened by a powerful alliance of Japanese trading companies, merchants and local politicians, who continue to push logging operations ever deeper into the interior.

Ngau, now 30, became concerned about logging in the late 1970s when its devastating effects began to become apparent. In 1982 he set up a branch of Friends of the Earth in Sarawak to help preserve the forests the Penans call "our bank and our shops." Ngau and his colleagues became investigators, exposing links between logging companies and politicians. Later, when the Penans found the courts stacked in favor of timber interests, they took the desperate step of blockading logging roads. Ngau and Friends of the Earth provided legal help and made the Penans' plight the focus of international protests. "It is our time to look after our place so that it will have a future," says Ngau, who spent 60 days in prison for his efforts to help the natives.

In the face of indomitable natives and pressure from foreign environmentalists, the Sarawak government has begun a dialogue with the Penans, and Malaysians have begun to respect those natives who choose to live in the forests. Thanks to Ngau and his colleagues, there is a sliver of hope that the grim sacking of Sarawak may be halted.

Following this honor, Harrison stood for and won the Baram seat in the Malaysia legislature in October 1990. As we greet him, he is cleaning out his office in order to sever ties with SAM while in government and move to Kuala Lumpur on the mainland for the parliamentary session. His election was a huge event in the Baram; he was native, an independent, and a well-known opposition activist with almost no campaign budget. He won five to four over the second candidate, who was well supported by the ruling party; a third was far behind.

The government and the timber companies have been understandably worried by Ngau's election. Natives are 45 percent of the population; most are not politically active, unlike the highly organized Chinese and Malay. In this case an opposition native won a seat hands down, with little professional backing or money. How many more might run and be elected, threatening the Malay hold on power, exciting the populace? Now the government and its newspapers regularly lash out at Harrison. The *Sarawak Tribune* of January 7, 1991, devoted an extraordinary eight columns to a direct attack on Ngau (entitled "Who Used Who?"), the gist of which is that he serves first world environmental/colonial powers (especially Prince Charles of Britain, who has spoken out against Sarawak logging). These first world powers, according to the *Tribune,* wish to strip Malaysia of sovereignty: "Local environmentalists merely echoed their master's voice. . . . The Penan issue became the surgical knife to cut the jugular vein of Sarawak's revenue, dissect the political stability, and slit the executive, judiciary and legislature." When one can be jailed without charges under the Internal Security Act, and has been, such talk from the establishment paper is troubling.

Harrison is a Kayan from Long Kesseh, about three hours by express up the Baram. Like most of the older students there he went to middle and high school at Long San and Long Lama (as did Richard) and learned excellent English. He speaks far better English than most Japanese students at Tokyo universities. After high school, he took a job in Miri, and while there he met Evelyne Hong, who was working on her classic book, *Natives of Sarawak*. Harrison became convinced that his people were in great trouble, and that they could be helped. Hong and her organization, the Consumers Association of Penang (on the Malaysian mainland, and perhaps the best activist organization in Southeast Asia), helped secure Friends of the Earth funding for Harrison to set up the SAM office in Marudi, in 1982. They began investigating logging practices. The horror stories began to filter to his office. They became a flood.

Harrison could hardly be less threatening, yet in his presence Richard seems hangdog; they speak in Kenyah, Harrison solicitous. Later I discover that Richard was embarrassed because he had worked for a timber company, and, to tell the truth, hoped very much to work for another. Harrison assured him that wages were no disgrace, and that he could be useful up the Baram if he wished. Richard said he wished, and was told the names of main SAM contacts in the middle and upper Baram. At that we all gathered around the map, Richard and I to hear news of the Baram, Jewin for a last review before Yokohama.

Harrison points to the large map of the Baram and the Tutoh districts, on the wall. We follow the Baram down from its source in the highlands at the upper right of the map, near Bario. The Baram begins as a stream flowing south across the Bario highlands, drops off the plateau and comes to Lio Matoh, the farthest up canoes can navigate. A river now, it flows southwest through the "upper Baram," a stretch of eight longhouses from Lio Matoh at the top down to Long Jeeh, at the big bend. There it turns suddenly north to become the "middle Baram," a stretch of seven longhouses from Long Palai and Long Anap upstream, down to Long San. Samling's logging has reached the middle Baram, and the upper Baram is being prepared. Below Long San the river becomes simply "the Baram," still far above Marudi, not to mention far above the lower stretch of the river down to Miri. The express boats — that is, scheduled transportation — go up as far as Long Na'ah, just below the middle Baram. There the first serious rapids discourage the big boats, so the middle and upper Baram may remain slightly isolated a little longer. The river miles in the middle and upper Baram are about equal. Both districts are way up in hill country, some peaks rising to five thousand feet, steep and forested. The middle and upper Baram are one or two or three days above Marudi, with increasing rapids as you ascend and above Long San, with virgin forest.

Harrison points out the present situation. The farthest timber camp up is Lambir, working a small concession above Long San in the middle Baram. Soon Lambir, a small, independent operator, will be bought out or merged with the huge Samling Timber Company, which has the rest of the Baram concession. Samling has bulldozed roads through the middle Baram and is cutting the big trees. Now they want an agreement with the upper Baram to begin work on roads all the way up to the highlands. Two roads are involved, one parallel to the river and right now turning the outside of the big bend, and one on the inside of the bend, slicing from Long San up toward Lio Matoh. Crucial to this second road is the bridge over the Baram at Long San, scheduled to open in June 1991. Then a lot of heavy equipment can come up. At the moment, just a few dozers and trucks have crossed the river at road's end, in low water or on barges.

Both Samling roads, the one outside the bend and the one heading cross-country inside the bend, are already at the edge of upper Baram native lands without logging agreements. The inside road has stopped.

The meeting at Long Moh, Harrison explains, between Samling and the upper Baram, will be crucial. The roads are already being held up; Samling wants a timber agreement. The meeting will be in mid-December, says the SAM office. But on that point, Richard and I will find out in a few days, SAM is wrong.

Our family checks into the Grand Hotel in Marudi for two nights. Linda and Sarah will buy their last souvenirs and return to America, and I want to read more Sarawak history. Besides, Juliette and I need police permits to go up the Baram to the edge of the logging. Jewin goes off with friends and Richard goes to the Penan house "Uma Sakai" near Fort Hose. These are his nights in Marudi with buddies. We don't even want to hear about it, and we don't.

What can I tell you about the Grand Hotel? After days upriver in longhouses, the shower and toilet, the air-conditioning, the privacy are delightful. Other things are more puzzling: why do the ants on the wall and ceiling go away at sundown, when the lizard (a gecko) comes out?

Does he eat ants? If not, why do the ants quit at six-thirty when we're eating crackers and drinking beer? But this is a minor conundrum, a naturalist's diversion, nothing like the metaphysical puzzles to come. After Chinese dinner "downtown" (fried rice with pork, egg, and *paku,* delicious Borneo ferns) we read for a few hours, then turn on the TV and settle back. "Let's run away with the Rich and Famous" says the tube, and we stare in disbelief as Robin Leach grins his maniacal grin over central Borneo. "Here's Barbara Eden's spa in California—Never say quit, Barbara. . . . Or perhaps you prefer the exclusive Winthrop Hills in Connecticut at $3,000 a week." After the relief of a tommy-gun ad for Dick Tracy Pizza at Pizza Hut, and for Fashion TV ("fashions with flair, fashions that dare"—surely this is coming across the border from the well-to-do Sultan of Brunei), we are back to Sheryl Lee Ralph of "LA Law": "Listen up baby, you say you want Paradise?" Off to a spa in Jamaica.

How can we be watching this, here in Marudi, where thirty years ago was still the isolated government post Somerset Maugham described in "The Outstation," where sixty years ago Tom Harrisson, with the blessings of the third Rajah, came with the first Oxford exploring expedition upriver of Marudi, where natives had not seen white men? Now the children of those natives are watching "Lifestyles of the Rich and Famous."

A bit boggled, we light a mosquito coil, wish the absent Richard luck without specifying its form—the spa has many faces—bid goodnight to man's best friend (the gecko), and go to sleep. In my dreams, I am boggled yet again as my history books come to life; I am haunted by Ranees, Margaret and Sylvia, those exotic exfoliations of the oddest colonial growth in Southeast Asia: Rajah Brooke.

CHAPTER 3 ❧
RAJAHS AND RANEES

Sylvia Brett's adventures began on that small green island halfway around the globe and far to the north. A letter was brought to her home in England, across from Windsor Great Forest:

Call it fate, that caused a letter to be sent to my mother from someone who lived four miles from our home, but whom she had never even seen. The writer explained that she was forming a small amateur orchestra in which she was very anxious my sister and I should join. It was signed MARGARET, RANEE OF SARAWAK. . . .

My sister and I scampered to our rooms and tore down every book of reference we could find, [and] seated there by the fire, Doll read me the story of the first White Rajah.

James Brooke was born in the European suburb of Benares [India] called Secrore, on 29 April 1803. He was the second son and favourite child of one Thomas Brooke, handsome and headstrong, and his parents doted on him. At the age of sixteen he obtained an Ensign's commission in the 6th Madras Infantry, and later he fought and was badly wounded in the First Burmese War. He returned to his family at Combe Grove, Bath, an invalid; so he sent in his resignation, and went on a voyage to China. It was then that he saw for the first time the islands of the Asiatic archipelago. To James Brooke they seemed an open invitation to adventure. . . .

In 1833 Thomas Brooke died, leaving £30,000 to his son, and James purchased a vessel of his own, the *Royalist,* the famous schooner of one hundred and forty tons that was to be the first British yacht to dip her inquisitive forefoot into Sarawak waters.

It was while he was in Singapore that James Brooke first heard of the rebellion in Borneo. The ruler, a Malay prince named Raja Muda Hassim, was desperate, and incapable of restoring law and order.

All that was known about Borneo at that time was that it was in-
fested by Dyak head-hunters, and by pirates who roamed the coast,
destroying native trade and terrifying the people. To James Brooke
it seemed a heaven-sent opportunity. "God has made me," he said,
"to be the suppressor of head-hunting and slavery in Sarawak."

So in 1840, with a handful of Englishmen, a few native boatmen
and little else but his cutlass and one muzzle-loading gun, James
Brooke landed in Kuching. He told Raja Muda Hassim that he was
confident he could put down the rebellion; but only on condition
that he was made sole leader in place of the unscrupulous Governor. . . .

In an agony of doubt and fear, Raja Muda Hassim at last reached
a decision, and made his famous declaration:

"If only you will remain," he cried, "I will give you all my country.
I will give you my government and my trade. All these things you
can have, and your generation after you, if only you will not desert
me in my hour of need!"

This was enough for James Brooke. With his tiny force he suppressed the rebellion, and returned in triumph to Kuching, where
he was received with acclamation and treated as a god. A year later,
on 24 September 1841, Raja Muda Hassim fulfilled his promise and
made James Brooke, then thirty-eight years old, the first White
Rajah of Sarawak.

So began the unique personal rule of the Brooke family in Borneo, which James maintained until his death in 1868, and handed on
to his nephew, Charles.

It was from the wife of Charles Brooke, the second White Rajah,
then living apart from her husband at Ascot, that my mother had
received the letter.

When my sister stopped reading, we sat and stared at one another.
We had been quite carried away by the intensely romantic tale of
pirates and head-hunters and a man who had been made a king.
Moreover, it was true. The Ranee Margaret of Sarawak was fact, not
fiction; and we were going to meet her.

I shall never forget our entry into Grey Friars. The Ranee Margaret was seated regally in a high-backed chair, talking about herself
and her three sons. . . . [S]he could dominate a room with her per-
sonality and her magnetic eyes, and enchant everyone in it. . . .

The week held only one day for us—the day we pedalled our way to South Ascot. In the hall of her house there was a huge macaw with scarlet wings, a terrifying bird, which was tied by a chain to a stand. When this strange woman was not seated at her piano she would recline in a blue armchair with the green parrot perched upon her wrist.

My sister played the side drums extremely well. I had no special talent, and with some apprehension chose the big drum, symbols, and triangle. No sooner had we joined the Grey Friars orchestra than we began to rehearse for a charity performance of "His Excellency, the Governor," with the Ranee's youngest son in the star part. . . .

When the rehearsal was over, the Ranee Margaret would ask my sister and myself to tea and then she would often talk about her life in Sarawak. She was estranged from her husband but had lived many years with him in the country he ruled on the north-west coast of Borneo. She had had three children, two boys and a girl, but they had died in their infancy of cholera and had been buried at sea. Because of the need for a son and heir, she had returned to the cold indifferent man she did not love and had borne him three more sons. . . . Their marriage had ended when he destroyed her pet doves and served them in a pie for her supper. We listened spell-bound to her stories; and I can see my sister, Doll, her great dark eyes fixed on the Ranee's face like a girl in love.

It was not until the Ranee's eldest son came home on leave from Sarawak that her scheme began to be revealed. Here were three shy, unapproachable, unmarried men who refused to go out anywhere or see anyone: and an inheritance which made it essential for them to marry. Only a clever woman could have found the solution to this problem in an orchestra composed entirely of desperate and willing virgins, duly assembled twice a week for her sons to look over. Blondes and brunettes, all shapes and sizes, nice, healthy, simple, unsullied, unspoiled girls. Not notably ornamental: in fact, a more homely group could hardly be imagined: and yet, in the fullness of time, each one of her sons became engaged to a girl in the Grey Friars orchestra.

You could not have found three more charming men than Vyner, Bertram, and Harry Brooke. Harry, the youngest, had his mother's charm and easy warmth of manner. Bertram, or "Adeh" as he was called, was a little diffident and constrained and covered with confusion when addressed. He had almost foreign manners, a reflection of his younger days in Heidelberg. Vyner was extremely handsome, with smooth fair skin tanned a rich brown by the tropical sun. He had the fine Brooke nose that nature has perpetuated in the outline of the Matang mountain in Sarawak; but he was so shy you felt that if you turned and spoke to him, he would rush headlong from the room.

I cannot remember when I first became aware of Vyner Brooke, of those startling blue eyes watching me across the room. I was too busy with my deafening drum and clashing cymbals. But, one day, he drew up a chair and sat beside me. I heard his quiet voice say, "I want you to let me tune your drum for you. May I, Miss Sylvia?" Nineteen hostile faces glared at me and nineteen disappointed hearts wished I would drop dead.

Sylvia became Lady Brooke, the second Ranee of Sarawak, and the last.

Sylvia and Doll Brett had certainly grown up in the privileged circles of England. Sylvia recalled one occasion: "The Royal children had managed to drag our big wagonette into the yard. Perched upon the driver's seat was Prince Edward holding the reins and cracking a long whip. Harnessed into the shafts were Princess Mary and Prince Albert, prancing, kicking, and neighing as loudly as they could. My father shouted to Prince Edward to get down at once. . . ." Yet for some reason, out of their family's association with the rich and famous, the children developed a talent for the primitive:

The time had now come for my sister Doll to be presented at Court. She had a soft delicate face like my father's, and her high-piled hair had turned from gold to a deep auburn. Outwardly she was intensely feminine, but it soon became apparent that this elegant façade concealed a masculine independence of spirit. She scorned the ritual of match-making, snubbed her escorts, and as soon as the round of parties was over, cut off her hair, dressed like a boy, and became one of the best-known figures in the painters' pubs of Chelsea. She learned to draw at the Slade School of Art, and eventually she packed up her belongings and went off to Taos, New Mexico with D. H. Lawrence. There she became one of the finest painters of Red Indian portraits and of pictures of their famous fire-dance.

Sylvia, not to be outdone by the dashing Doll gone primitive in Taos, became a Ranee. From the beginning of her life to the end, she was a maverick. Before her marriage to Vyner, she had already published two volumes of short stories (praised by her friends J. M. Barrie and George Bernard Shaw). Later, she told Errol Flynn to his face, over dinner, how he had destroyed her synopsis by writing a fantastic script about James Brooke called "The White Rajah." In the 1950s she was on the skids for years, alone, in Hollywood (where one of her daughters lived briefly with a professional wrestler), drinking and broke in Barbados and New York. Hers is an extraordinary tale of colonial adventure and aristocratic demise, told with no little irony and humor.

The story she tells is not only of inheriting an empire, but of giving it away. As the Japanese advanced in the Pacific in 1941, Vyner and Sylvia fled to Australia. After World War II, Vyner ceded Sarawak to the British government. In 1963, Sarawak and neighboring Sabah joined the new Malaysia. Thus colonial geo-eccentricities were perpetuated; old Dutch south Borneo went to once-Dutch Indonesia, and British-dominated north Borneo went to once-English Malaysia. Brunei, unlike Sabah and Sarawak, opted for independence. This introduces the very obvious question, Why isn't Sarawak, or Borneo, on its own? The question is still alive: in January 1991, the Chief Minister of Sabah was arrested by Malaysia for allegedly plotting secession.

It would be naive to think that significant foreign influences have only recently come to Sarawak, that British rajahs and ranees, Malaysian ministers, Japanese companies, or European environmentalists are corrupting "innocent natives." By the eleventh century, Sarawak had significant trade with China, Indochina, Indonesia, and India. By the twelfth century, Hindus had come from Madras (archaeologists have found Gaja remnants, Tantric remains, and a bull from a Shiva temple, near Kuching). At least since the fifteenth century, Tang and Ming Chinese jars for storage, ceremony, and prestige have found their way up to the Bario highlands, an incredible journey by boat and foot for a 150-pound ceramic jar with red dragons. The jars are there to this day, prized family heirlooms. In 1944, Tom Harrisson was living in the highlands at Bario, an adventurer and a British officer directing the resistance. (Later, he became curator of the Sarawak Museum.) He made discreet inquiries concerning the value of the heirloom Chinese jars. He noted the "intricacy, elasticity, and apparent illogicality" of their worth in trade:

Four ordinary dragon jars = one male buffalo calf.
One well-grown male buffalo = thirty yellow glass (= 'bone') beads (worn in the front of women's caps).
Five buffaloes, five fat pigs, three hump-back bulls, two goats, two ordinary jars, two small jars, two gongs, two fine parang knives, ten mats, ten fish nets, ten fowls, ten Pa Mada pots, ten rolls of best leaf tobacco, one hundred yellow cane beads and two hundred [three ounce packages] salt = one old dragon jar of red-bodied stoneware (if you can get it).
One old dragon jar = one human life.

In the poorest longhouse kitchen, a five- or ten-gallon Chinese trade jug is still the preferred water vessel, keeping the kitchen water cool and clean under a wooden lid, to be ladled out on demand. Today when Penan come out of the forest with baskets, parangs, and blowpipes for the local and tourist trade, to exchange for lamps or fuel or rice or clothes or cash, they are following old trade patterns.

Throughout the nineteenth century, Chinese came to Sarawak to settle, many of them Hokkien from South China, seeking gold, diamonds, antimony, or tin, or just adequate land—very similar to settlers in the American West. Always they traded and developed retail businesses, kept ties to their homeland and language, built their own schools, and created an international presence in the mining camps and coastal towns.

In his journal of 1839, James Brooke called Borneo "a trader's paradise." His lists included "antimony, timber, Malacca cane, rattan, beeswax, bird's nest for the Chinese epicures, pipe clay, rice, sago, vegetable tallow, perhaps gold and diamonds." By 1910 there were oil wells at Miri, and now oil and natural gas are important. Alas, the best fields lie just across the border, in Brunei, which rivals Kuwait as the richest kingdom on earth.

Rubber has been the main cash crop. High prices and expanded plantations after the First World War gave way to a disastrous fall in 1929; many planters lost everything in the thirties. Large-scale plantation development, which requires controlling and clearing the land, was never vigorously pushed by the Sarawak Rajahs. Compared to mainland Malaysia (under British rule), the plantation system in Sarawak was more of a family farm system. Now, however, with native land rights weakened, logging is sometimes followed by large oil palm, tea, cocoa, or rubber plantations, where natives work for wages.

The three Brooke eras (James 1841–68, Charles 1868–1917, and Vyner 1917–46) constitute one of the strangest centuries of governance in history. First of all, the Crown never knew just what to do with Rajah James, who for his part remained, as Robert Payne has observed, "baffled by the British Government, which seemed to regard him simultaneously as an honorable and cherished ally and as an interloper, who had the audacity to found a colony without asking permission from London."

When Charles, as Second Rajah, came to London to be presented to Queen Victoria, he was informed that his audience would be as a private gentleman, "Mr. Brooke." However, they placed in brackets after his name: [Rajah of Sarawak].

The bracketed Rajah didn't much care how they did it in England. He came to the mother country, of course, for his son's wedding to Sylvia. As Sylvia tells it,

> the old Rajah [Charles] . . . had been dragged unwillingly to the marriage of his son [Vyner], didn't know who his host was, loathed the whole affair, and only wanted to leave as soon as possible. He turned to the first man he saw, and said, "How the hell can I get out of this damned house?" The man happened to be my father, who was so astonished that he meekly showed him the door.

The Brooke rule was especially odd, however, because all three men were so captivated by the country they had captured. In the battle of imagination, Sarawak can be said to have won, or at least to have fought to a draw. The three men were not only uncomfortable in England, they were often uncomfortable in Kuching, and happiest in the longhouses or raiding obstreperous tribes. One thinks of the British clubs in India, so punctiliously British in the face of foreign customs; in contrast, Sylvia remembers manners on the porch of the Astana palace in Kuching:

> Once, in a lull in the conversation I said in my loudest and clearest voice, "Listen, it's begun to rain." There was a stricken silence, and turning my head I saw the old autocrat, ruler of fifty thousand square miles and a half a million people, standing up quite unashamedly before his embarrassed guests and watering the cannas over the veranda rail.

Retired Brooke officers recall the Sarawak style:

> "The Brooke tradition was very firm," declares Robert Nicholl. . . . "Every officer who travelled lived with the people. There was no question here of setting up half-a-dozen tents and dressing for dinner in the jungle as the great proconsuls did in Africa. That was out as far as the Brookes were concerned. You stayed in the longhouse and you lived with the people." The difference between Sarawak and Malaya, as Bob Snelus saw it, was that "we behaved like natives. We accepted Dayak [native] conditions as they were. I always insisted on having my bath in the river first and then got into my *sarong* and *baju*."

The preference for longhouse style extended to economics; Sylvia observed of old Charles, whom she did not like:

> Whatever may be said of the Second White Rajah, his people knew that he had dedicated his life to them and that he kept faith with them. He would not allow commercial or industrial developments, but encouraged and extended agriculture and education. . . . He never forgot the fundamental dictum: "Sarawak belongs to the Malays, the Sea Dyaks, the Land Dyaks, the Kayans and all the other tribes, and not to us. It is for them we labour, not for ourselves."

Charles, however, did not mention the Chinese. All three Brookes were suspicious of Chinese and European enterprise, development, and industrial progress, for which the Chinese still have not forgiven them. "They kept Sarawak down; now it is finally moving" is a common sentiment among Chinese entrepreneurs in Kuching.

In addition, James was seriously prejudiced against the Chinese before he ever came to Borneo — in Singapore where he saw them exploiting the Malays, and in Canton. He disliked them physically, morally, and intellectually. It was the Malay and native life the Brookes admired, in a paternal fashion. In his first visit in 1839, James Brooke thought the natives honest, kind, tactful, fearless, and industrious, with a natural nobility. But the Chinese he feared (and indeed he barely survived their armed rebellion in 1857). A Chinese settlement only five months' old had acres of rice, betelnut, corn, and potatoes, and they were mining antimony, diamonds, and hoped for tin; James wrote in his diary, "The race are worthy of attention, as the future possessors of Borneo."

In their distrust of development and their love of land, the Brookes might be called Jeffersonian agrarians. Jefferson had opposed manufacturing in America ("let our workshops remain in Europe") because he had seen the terrible working conditions in the English cotton mills. He and Adams and others thought that working for wages was debilitating, and that a country, and the vote, should be placed on the free shoulders of men who managed their own land. Coming back to Asia just a few years after Jefferson's death, and during the time of Marx, James Brooke shared some of his period's fear of its own industrial progress. A former Brooke official sums it up:

> It was the policy of the Rajahs, First, Second, and Third, not to develop the country. They would not have European companies in to plant rubber all over the place and mine it and work the timber. They wished that the local people should plant their own rubber gardens and work their own timber, or, if they wanted to work gold, to go and work it. They weren't wage slaves; they were free to earn their own living. And this is a very important point in Brooke rule, because it was almost the only country where such a thing was done. The result was that nobody was rich and nobody was poor.

One begins to imagine London's concern; peeing off balconies was one thing, but such views of appropriate economies and of native land rights . . . well, these Rajahs sounded antiprogressive, damn close to anticapitalist. The Brookes themselves realized the conclusions suggested by their own tendencies of style and thought. In 1907 Charles foretold the fall of colonialism. In a pamphlet printed in London, *Queries, Past, Present and Future,* he predicted that "before we reach the middle of this century all nations now holding large Colonial possessions will have met with very severe reverses." He had no doubt why:

"The fault lies in ourselves as governors, and mostly arises through the want of care and of knowledge of the native mind." On the Indian mutiny, he commented: "The people were never questioned as to their wishes, and later on, when they demur or show forcible opposition, they are called rebels and are shot down."

Following such lines of thought, a colonial government might well self-destruct, and that is what Charles's son Vyner finally considered appropriate in 1946. He thought the Rajahs obsolete. Before the Second World War, he had turned power over to a native council. After the Japanese occupation destroyed what little economy and administrative infrastructure Sarawak had, Vyner thought Sarawak needed Britain's capital and protection. When he declared his intention to hand Sarawak over to England as a colony (working toward independence within the Commonwealth), many defenders of the Brooke rule understandably came forth. Challenged on the wisdom of his decision, Vyner attacked his own family dynasty in a scathing answer to an article in the London *Sunday Times*. He concluded:

> In all the criticisms which have appeared in Parliament and Press I have sought in vain for a single constructive suggestion which might contribute towards the establishment of a happier and brighter future for the natives. All that is urged is a return to "the good old days." But those days were good only for the British residents and not for the natives.

The natives might have suffered less from benevolent Brooke despots and a benign neglect than they would have in other hands, yet still the time for the odd family enterprise had come to an end. The Japanese occupation had broken the back of Brooke rule. For better or worse, in one year Sarawak was liberated from the Emperor of Japan and the Rajahs of England.

After the demise of Brooke rule, Sarawak had a strange, seventeen-year interregnum. There was considerable opposition to being a British colony, and the second colonial governor, an Englishman, was stabbed to death. In the late 1950s, a growing Indonesia, under Sukarno, was poised to expand from south Borneo into north Borneo, while the Philippines were eyeing Sabah, historically close to them and their people (a chain of islands stretches from Mindanao to northeastern Borneo; it is still a no-man's archipelago of squatters, smugglers, pirates, and insurgents). In addition, after the communist revolution in China, there were various Chinese communist guerrilla movements in both Borneo and Malaysia. Then in 1962, a revolt in neighboring Brunei sought to kick the British out of all north Borneo, and gain independence for Sarawak, Brunei, and Sabah. At this time Malaya, the mainland state which Britain had granted independence in 1957, proposed a new federation including all of north Borneo. Sarawak and Sabah, unstable and possibly threatened from without, joined; Brunei stayed British and was granted independence in 1984.

In 1963, Indonesia felt threatened by the new and growing Malaysia to its north, though Malaysians would say their federation was formed partly to counter the Indonesian threat. Whatever the sequence, Sukarno began aiding the Chinese communist guerrillas in Borneo at the Sarawak-Kalimantan border, where the Chinese rebellion had begun in 1857. From 1963 to 1966 a "Confrontation" continued between Malaysia and Indonesia. The situation was violent though never a declared war; British planters on the mainland were under siege by communist and Malaysian liberators of the land, and various local guerrilla groups operated on both Borneo and the mainland. Finally Indonesia changed course in 1966, the Borneo communist guerrillas had to fight from Sarawak bases, and in 1973 they signed a truce with the government. Many Chinese intellectuals, opposition leaders, or even establishment ministers (such as James Wong) have been jailed at various times, on charges such as fomenting rebellion, plotting secession, exciting ethnic (Chinese) passions, or, in the case of both James Wong, government minister, and Harrison Ngau, native activist, on no charges at all, under the Internal Security Act.

After the 1960s, Malaysia gradually became the most progressive country in Southeast Asia; the mainland, relatively industrial, clean, and efficient, is not a "third world" country.

To this day, the Malaysian nation is a fascinating experiment in multiracial governance, but with crucial differences from the United States. Assimilation is not so pervasive an ideal in Malaysia; schools, communities, and political parties follow ethnic lines or are obvious "coalitions." Industries are not to be "color-blind," they are to hire ethnic groups at each level in numbers approximately proportional to the makeup of the general population. The "pan-Asian face," a generic Asian look favored by models, was banned from television in 1988;

models were to be identifiably Malay, or Chinese, or Indian, or native. The Malays are a 51 percent majority in Malaysia and came there one to two thousand years ago. As *bumiputra,* sons of the soil, Malays dominate politics and have many legal advantages—as in having quotas for colleges (which otherwise would be thoroughly Chinese), in starting businesses, in obtaining loans and licenses, and in holding land. The talented and industrious Chinese minority is clearly and legally and mildly suppressed, so that the Malays can "catch up." Whether they have caught up, or ever will, is hotly debated—in private. There are laws against discussing race in relation to politics. In 1987 some leading Chinese politicians were detained for allegedly exciting ethnic feeling. They were all members of an opposition party.

The national language, Bahasa Malaysia, serves as a unifying force, but upriver in Borneo you will not elicit warm smiles by attempting to use it. To some extent, it is the language of the oppressor. Often you will be corrected and told the word in Iban, or Kayan, or Penan.

Sarawak and Sabah entered the Malay union with special provisions; they are both on their own, yet are states of Malaysia. The Sarawak population is 45 percent native, 30 percent Chinese, and 20 percent Malay. The Sarawak top ministers are almost all Malay, and the timber concessions are Malay and Chinese (and a few wealthy natives). Logging companies (like most businesses in Malaysia) are Chinese owned. The nation—especially the mainland—is moving forward very well, with less destructive booming and corruption than Thailand, less tyranny than Indonesia, less poverty than anywhere in Southeast Asia outside of (Chinese) Singapore. However, in Sarawak a Chinese opposition party has been formed, and in Sabah a native-Chinese opposition coalition swept the elections against the national Malay-Chinese party. Let us say that ethnic backgrounds are strong and valued in Malaysia; they are not even supposed to disappear. They are written into law and make the politics quite interesting.

One thing, however, is clear: since the end of Brooke rule in 1946, and especially since joining the Malay federation in 1963, Sarawak has been thrown open to development by British changes in the land laws, by Malay ministers, by Chinese entrepreneurs, and by enormous outside capital; and the natives have lost control of the lands on which their ancestors made their livings. When he was an old man, Charles spoke to his native council in Kuching in 1915:

I beg that you will listen to what I have to say, and that you will recollect my words, and endeavour to call them to mind to my Successor. . . . I have lived in this country now for over sixty years, and for the greater part of that time as Rajah. . . .

Has it ever occurred to you that after my time out here others may appear with soft and smiling countenances, to deprive you of what is solemnly your right—and that is, the very land on which you live, the source of your income, the food even of your mouths? If this is once lost to you, no amount of money could recover it. That is why the cultivation of your own land by yourselves, or by those who live in the country, is important to you now. Cultivation by strangers, means by those who might carry the value of their products out of the country to enrich their shareholders. Such products should be realized by your own industries, and for your own benefits. Unless you follow this advice you will lose your birthright, which will be taken from you by strangers and speculators who will, in their turn, become masters and owners whilst you yourselves, you people of the soil, will be thrown aside and become nothing but coolies and outcasts of the island.

This is exactly the situation upstream in the Baram now, the situation Fujino called "difficult." Charles once again showed great prescience in seeing the dangers of economic development by colonialists, whether they come from abroad or from Kuching. The forests are cut, plantations move in, natives who were once free and independent on their own tribal lands become wage slaves for absentee landlords. Sylvia said her husband, Vyner, finally ceded Sarawak because "he would rather make the King of England his heir than throw Sarawak to the commercial wolves who would devour it." Unfortunately, the dangers lay within as well as without.

The Ranee Margaret, though she could no longer stand the company of the autocratic Charles and had to leave the Sarawak she loved, nevertheless paid tribute to the Brooke treatment of the natives:

Being monarch of all he surveyed, unfettered by tradition, and owning no obedience to the red-tapeism of Europe, Rajah Brooke laid the foundations of one of the most original and, so far as justice goes, one of the most successful Governments that perhaps has ever been known, its most salient feature being that from its very beginning the natives of the place were represented by their own people and had the right to vote for and against any law that was made by their Government.

In Marudi, the next day, we walk out of the Grand Hotel into blinding sunlight. We stroll the bustling three blocks from the hotel to the dock, to put Linda and Sarah on the express downstream, loaded with sarongs and parangs and Penan baskets and a blowpipe. Along with my history books, which I have finished. Richard appears for this ceremony. The handshakes are touching, as they can be among affectionate strangers who have felt their common humanity yet know they will never meet again. The express toots three times, backs out banging into the other boats, and heads full throttle downriver, dodging rafts of logs, toward the outside world. We wave, Japanese style, until they are out of sight, to make sure we will see them again. They are going back to America. As I look at Richard, I cannot guess his idea of where they are going, of what America, or Japan, might be to him — the seat of power, of economic empire, of legendary riches, of Mutant Ninja Turtles, "Dallas."

Richard walks with us up the hill to Fort Hose, named for the great naturalist and good friend of Ranee Margaret, that indomitable woman who late in life and estranged from Charles, living in Genoa, Italy, took Oscar Wilde's wife and sons into her home when no one else would have them. We think of Victorian style and British mavericks, of Brooke officers who once lived here on this little, grassy knoll with a white "fort." Not so long ago. When I was born, the third Rajah ruled.

I remember Somerset Maugham's story "The Outstation," set in Marudi of the 1920s. A young English gentleman, fallen on hard times, had to leave London and join the Foreign Service:

The night before he sailed he dined for the last time at his club.

"I hear you're going away, Warburton," the old Duke of Hereford said to him.

"Yes, I'm going to Borneo."

"Good God, what are you going there for?"

"Oh, I'm broke."

"Are you? I'm sorry. Well, let us know when you come back. I hope you have a good time."

"Oh, yes. Lots of shooting, you know."

After a few years in Marudi, Warburton "was accustomed to loneliness. During the war he had not seen an English face for three years." There was not another white man for two hundred miles. The Malay and native servant boys in the officer's bungalow, at suppertime, must have felt the same distance and wonder that native boys feel now watching "Dallas":

> Mr. Warburton flattered himself that he had the best cook, a Chinese, in Borneo, and he took great trouble to have as good food as in the difficult circumstances was possible. He exercised much ingenuity in making the best of his materials.
>
> "Would you care to look at the menu?" he said, handing it to Cooper.
>
> It was written in French and the dishes had resounding names. They were waited on by the two boys. In opposite corners of the room two more waved immense fans, and so gave movement to the sultry air. The fare was sumptuous and the champagne excellent.
>
> "Do you do yourself like this every day?" said Cooper.
>
> Mr. Warburton gave the menu a careless glance.
>
> "I have not noticed that the dinner is any different from usual," he said. "I eat very little myself, but I make a point of having a proper dinner served to me every night. It keeps the cook in practice and it's good discipline for the boys."

I look at Richard's smooth, strong back and calves as he walks ahead of us, clad only in shorts. A paddler's shoulders, a hunter's legs. Are we any less strange to him than Warburton was to those Malay boys? Our beautiful, single daughter and niece are traveling about alone, one a student, the other a lawyer. We say we're just schoolteachers, but we fly to his country and rent longboats and him at our pleasure. He doesn't come to America. For all our democratic pretenses and open manner, are we not more puzzling than a Rajah's agent acting exactly like a Rajah's agent? We come from the kinds of places that are sucking up whole forests, his forests, and yet we seem so friendly, and say we're against it all. We reach the top of the hill, leveled for the white frame "fort" building still in good repair, and look down where, says Maugham, the Rajah's agent used to look down:

> The fort was built on the top of a little hill and the garden ran down to the river's edge; on the bank was an arbour, and hither it was his

habit to come after dinner to smoke a cheroot. And often from the river that flowed below him a voice was heard, the voice of some Malay too timorous to venture into the light of day, and a complaint or an accusation was softly wafted to his ears, a piece of information was whispered to him or a useful hint, which otherwise would never have come into his official ken.

I look down at Marudi and the river, thinking of modern centers of power, of Malay ministers with Hong Kong bank accounts, of Japanese import firms worth billions, of Chinese timber companies cutting fifty acres an hour out of one camp. Whatever its virtues and vices, the Brooke world—it seems for the moment a kinder, gentler world—is gone.

At Fort Hose, Juliette and I are interviewed for police permits to visit the upper Baram. We are sent to the cordial, handsome, and articulate district supervisor of police, Mr. Wagner, a Malay. He says that the permits are for our own protection, so that in case of trouble they will know where we are. Do we have letters stating the purpose of our visit? We do, from the Sarawak Museum in Kuching, that most wonderful Brooke legacy, perhaps the best museum in Southeast Asia. Juliette is writing an article for the *Sarawak Museum Journal* on Penan adaptations of native dance in Sarawak. I am researching a book on Sarawak and Japan. Mr. Wagner's bookshelves are heavy with Maugham, Dickens, Tolstoy, Stendhal. He is genial, and no fool. He kept us waiting fifteen minutes, plenty of time to clear his desk, and yet in plain view, to one side of the desk, facing him, is the yellow cover of SAM's radical publication, *The Battle for Sarawak's Forests*. He pretends it is not there. I pretend it is not there. Is this a test? What message is he giving us?

We suspect the permits are to control and suppress visits to areas being logged. Still, this is the land of headhunters, and we are not absolutely certain who is protecting whom, and in the backs of our minds float strains of the upriver lullaby:

> Listen well, my little rice-basket,
> Grandfather's head hangs over the fire.
> Go and avenge us:
> Do not let us give you milk in vain.

CHAPTER 4
UPRIVER

November 13, 1990

The whistle echoes off the white walls of Marudi at 7:15 a.m. Richard
and Juliette and I drain our coffee and walk down the gangplank to the
pier. The girls are gone, and we are heading "deep inside." The silver
Long Lama express blasts again, and this time, rain threatening, we
take our seats down in the dark cabin. The music, at top volume: "This
is my island in the sun . . ."

"Upriver in Borneo." The phrase, to Juliette and me, has always
meant exotic adventure, excitement, romance. The boat vibrates with
the deafening diesel and we back out into the current, smashing into
the other express boats tied side by side at the wharf. We are off to
Long San, a bit beyond the express service, one or two days away depending on luck. As the gas pump, dock, and palm plantation slide by,
I find myself wondering what "upriver" means now—after the Rajahs,
after the Malays, after the logging of the last ten years.

Music magic—cassette science fiction, giving the impression of one
world—continues to pour from the express boat speakers:

> It was down in Louisiana
> Just about a mile from Texarkana
> In those old, cotton fields back home . . .

A four-year-old Penan girl in front of us, wide black eyes in a nut
brown face, long straight black hair, stands up in her seat, looks over
the back and discovers my beard. She is not afraid. If I return her
gaze she looks away, but for an hour I defeat television for this audience of one. My competition on the screen at the front of the cabin is
a wrestler named Honky Tonk, an Elvis imitator who looks mean but
may, it seems, have a heart of gold.

Honky Tonk wins his match, and in his interview, according to the commentator, will give us a surprising glimpse of his sideburns, and his humility. In the seat beside me, Richard takes a banana out of his bag. I ask him what sort of job he expects up the Baram, with Samling. He shrugs and with disdain says, "Pickup drivers are errand boys." He would rather be paid by the ton of wood processed, as are logging teams, drivers, boatmen. On the Tutoh he had begun by washing off muddy equipment in the yard, and had become a steady pickup driver, delivering men and fuel and supplies to various points up the roads which stretched half a day from the camp. But when the trees were gone and the road had stretched too far, it was time for the main camp to move up. That meant Richard would have to live in a timber camp, completely abandoning his tribal and family life, or give his job to someone in a longhouse upstream. Like most boys, when the logging had passed he was unemployed. But while Baya Timber was on its way through the area, he had gained a lot of experience. The boys fill in for each other at various jobs; as long as the job is done the company doesn't care, and the boys are experts in sorting out paychecks at the end of the month. Richard gained other experience too: along with many other natives from Long Bangan and Long Tarawan, he blockaded Baya Company's road in an effort to stop the logging. Police broke up the barricade. Back to work.

The logging companies have brought considerable amounts of money upstream, and by local standards the workers are well paid. This itself is a problem: the logging has brought in so much money so quickly, mainly to a few young pockets, that it tends to overwhelm— to make a mockery of—the local economy and culture. Here one can sympathize with the timber companies: the gap between Tokyo markets and upriver economies is so great that justice is impossible; if the company pays by local standards, they are exploiting the natives shamelessly; if they pay by standards commensurate with their trade on world markets, they are in many ways destroying a local subsistence economy. Such a gap between rich and poor, rich at least in cash and poor in cash, simply cannot be bridged without violent disruption of local patterns.

The question then becomes, is the new influx of cash stable, something to build on, or is it a drug, a temporary high which will leave the local area depleted and depressed? Since the timber companies themselves say there are only seven years left in Sarawak, one fears for the loss of local subsistence economies, which depend not only on a productive forest but on old relationships, habits, values, and skills — much of which can be lost in a generation. There is no transition period; a man and a woman know one way of life, which works pretty well, then suddenly their son is living in another world.

So, I ask again, to the son of parents I will meet in a week, what do you want to be? With a cocky grin, Richard says that he wants to be a cowboy. We laugh at our word. The logs in the river are herded by "cowboys" in little twelve-foot steel boats with five short spikes on the bow stem, and a Suzuki or Yamaha fifteen or thirty horse outboard. They roar back and forth rounding up logs by pushing them with the spiked bow. They slam them together to make rafts of twenty, fifty, seventy logs, then accompany the logs downstream, nudging the rafts about in the river, catching strays and tying them back in.

The boats are jockeyed recklessly, full throttle for fifteen feet, often riding up on the logs, ripping out wood with the front teeth, backing off fast before the boat can flip. The kids are crazy; the boats are fast and tough; it looks like a lot of fun. Finally the small rafts, herded by one or two cowboys through the middle Baram rapids, are delivered to wherever the river is gentle enough for a tug to pull the big rafts down to Miri.

Richard has a cousin (almost everyone up and downstream is a cousin) in Long San who wants to raft; they could go in together on a boat. The boat costs $380 U.S. (Malaysian ringgit are converted at 2.6 to the dollar, a November 1990 average. I have converted the round ringgit figures given me by many sources; therefore the dollar figures may seem oddly exact; "about $540" a month, etc.) The outboard costs $1,350 (a Suzuki 30; Richard says no slower horse will do), and you pay the company back as you take down logs. A raft might be sixty tons at the most, paying $3 per ton (usually split between two boats). In an excellent month one person might make three or four runs and earn about $400; and since good months (little rain on logging roads, yet river high enough) rarely follow each other, the optimal payback period for the motor and boat (without taking out wages) is more like six months. Then you make money, and own the boat. In practice, of course, most boys want immediate cash for living, so they pay back with half their wages, or less, over a longer time. After a year or two, you might own it all, yet need a new engine.

It is impossible to compare precisely expenses upriver with living in America. In some ways, Borneo is quite expensive; on the other hand, living with relatives in a self-sufficient longhouse you need little income, since you are almost outside of a cash economy. In longhouses, there is no restaurant, although up to Long San there is usually a mom and pop store. Nowhere above Marudi are there cars, bars, or movies; but you could buy a TV, videocassettes, and a player in Marudi and take them upriver to your home, because most longhouses have generators which run a few hours each evening. VCRs in Marudi cost the same as in the United States, slightly less than in Japan. From Marudi down, hotels and restaurants are fairly cheap. A nourishing plate of pork fried rice mixed with egg and vegetables is 75 cents in Marudi or Long Lama, a can of Carlsberg beer is $1.50. But those are big-time luxuries to any but townsfolk and timber workers. Gasoline, used for all river activities including fishing a mile upstream, costs $3 a gallon at Marudi and $5 a gallon near the top of the river, at Long Moh.

Longhouse rent is free, once a family fells a tree and saws it into planks and builds a room, called a "door," off the common porch. The family "owns their own home" within the longhouse, a condo if you will. Fruit trees and rice paddies are planted everywhere family by family, and many greens and fruits are gathered from gardens and from the forest until the logging comes through. You could live in a longhouse for almost nothing; on the other hand, wanting just a few things, you could spend quite a bit of cash. For instance, you live in Long Moh, near the top of the Baram, and want to spend a week going down to Marudi to buy something and come back. The gas round trip will cost you $300, without figuring wear on your motor. Not many can go down, and not very often, and what they bring back is precious. Remember that a family working rubber may earn $170 a month.

If you had grown up in a longhouse, the power and noise and excitement of being a cowboy, the prospect of owning your own boat, plus maybe $150 a month, would look pretty good—if you could live at home, which is any longhouse with a cousin.

Richard hopes (dreaming a little now) to get some experience on a dozer or log truck, and move up into real money. I ask how much. He says a log truck driver—his good friend's brother—can make $8,000 in a month. I sit back stunned; the Penan girl recoils, worried that I have turned to stone. I cross-examine Richard using pen and paper. In a dry month with loads of backlogged trees and good cutting weather, by sleeping two hours a night in the cab and working twenty-two hours a day with an occasional day of solid sleep (they are paid per ton), a truck driver—maybe a twenty-one-year-old Iban or Kenyah boy like Richard—might make $5,000 to $8,000 U.S. in one month.

"What can you do with that money in a longhouse?"

Richard shrugs and replies, "What can you do with $1,000, or $500? Put some in the bank, give some to relatives. Drink a lot."

"A lot? You can't drink that up."

"Sometimes you're paid once a month, sometimes once every two or three months. They go down to Miri, buy beer by the case—you can spend a lot of money drinking. I know guys who earn pretty good wages, say $300 or $400 a month, who drink it all."

"But $8,000. That's so much money . . ."

"Well the company takes 40 percent for taxes, but yeah, there's a lot of money around. Last week the head of Citibank in Hong Kong flew in to Long San with his wife—separate helicopters—to look over his investments in the Baram. They landed at the strip at Long Akah and zipped up to the mission soccer field at Long San. Just a couple of days ago. My friend at Marudi—he's from Long San—watched them land. He asked the priests who they were."

I think it is too much for me to comprehend, and sit back again. Why does all this money, the Citibank family in his and her helicopters, bother me so? Then I remember some documentary footage I saw in a Swain Wolfe film, footage borrowed from Smithsonian archives, of the first white men among a stone age tribe in New Guinea, not too far east of here. The old native man is showing the white fellow, who looks to be a sallow and paunchy anthropologist, how to light a fire. They are squatting on the ground, around a pile of kindling, and the old man's ten-year-old grandson is squatting too, watching every move. The old man strikes the flint again and again but cannot get the shavings going. Finally, the white man grunts, as if enough film has been wasted, rummages in his pocket, pulls out a cigarette lighter and reaches over without a word. Flick, the flame shoots out, the shavings are lit. The boy looks up at the white man, then at his grandfather, then, eyes narrowed, back to the white man. Freeze. An entirely new, undreamed of, and dominant power has entered his life. And it comes from someone else's father or grandfather, not from his own.

His world will never be the same. It is not just a matter of technology. Something like that is going on here, around me, with this influx of money and machines, this touching of cash to the kindling of young lives. I feel to the bottom of my heart that it breaks chains of respect and relation, and is a lie, for in no deep and lasting way are we more powerful men than the fathers of Borneo.

The flickering television at the front of the cabin attracts my attention. Pro wrestling is over, and a documentary has begun, the only documentary I will ever see on an express boat. It is about the Penan. The narration is Cantonese; the film was made in Hong Kong. The camera is lingering on topless Penan women, mostly old, beside a hut; then it is zooming in on their breasts, which fill the screen, pair by pair, and now the camera comes to rest on the pubic areas of naked children. The Chinese narrator is laughing. The little Penan girl on the seat in front of me is still staring at my beard. The camera keeps bulldozing its way through the people of the forest. I cannot bring myself to ask Richard, half Penan, what he thinks. I cannot ask myself what I think. I am sick of power, power of money, of helicopters, of cameras, of lighters, of the first world, of myself. I get up, and walk back past the nun in the blue habit reading the *Borneo Post,* past the *Playboy* poster on the back wall — the model's arms folded across her white breasts, beneath a knowing smirk. Breasts upriver are not vested with such power — why? Then I go up on deck to get some air.

We are passing the complex of timber camps at Temalah. From here you can drive on a brand new dirt logging road all the way over to Long Seridan on the Tutoh, where twenty years ago was "unexplored" jungle on European maps, where five years ago the Swiss renegade Bruno Manser hid out with the Penan, trying to help them defend the forest they had long called home. Temalah has three huge timber yards and docks in a row, about a mile and half of riverbank completely bulldozed, stacks of logs, outbuildings; we stop at one camp by nosing into the muddy bank; two men jump off with two cartons of fresh vegetables and another of Australian apples, Granny Smith Grade 1. Above Temalah the river has risen and everyone is floating their logs. We roar upstream past twenty rafts and their cowboys, each raft eighteen to forty logs, twenty rafts of about six hundred logs in all — probably four hundred trees — in the six minutes I am counting. Four hundred trees! At least another two hundred acres logged, neighboring trees smashed, vines torn. This is fun? Why didn't we go to Belize? But then, Central America has problems too.

At Marudi, when the raft of logs went by, I thought of the buffalo. Now, on the Long Lama express, watching raft after raft, I wonder if the comparison was apt.

It happened in Montana in the early 1880s, the extinction of the buffalo. There were a lot out West, maybe twenty or thirty million, maybe more, and most of them were still left in 1878. At one time they must have seemed as innumerable as the trees of Borneo. But after the Civil War the United States wanted the grass for the cattle. We wanted the cattle for the beef industry, largely a corporate enterprise, and we wanted to ruin the Indians' nomadic economy and get them off the land so it could be controlled for our profit. General Philip Sheridan told the Texas legislature in 1873 that the buffalo hunters "are destroying the Indians' commissary; and it is a well-known fact that an army losing its base of supplies is placed at a great disadvantage. . . . Send [the buffalo hunters] power and lead, if you will, but for the sake of lasting peace, let them kill, skin, and sell, until the buffalos are exterminated. Then your prairie can be covered with the speckled cattle, and the festive cowboy. . . ."

In my home of Montana, the slaughter came west with the railroad. By 1880 the trader Joe Kipp at Fort Conrad had heard rumors of buffalo depletion, and his friend Schultz had letters from the East asking what was going on. Rancher Granville Stuart that spring found rotting carcasses dotting the prairie along the railroad line for a hundred miles. By 1882, the Bozeman *Courier* was saying: "Friends of Montana everywhere are becoming seriously impressed with the fact that the large game of the country is being rapidly exterminated, and that a few years more of such wholesale slaughter . . . will result in the almost, if not total, extinction of most of the valuable species of game in the territory."

By 1883–84, "starvation winter" on the Indian calendar, the buffalo were gone. A homesteader in southeast Montana, Nannie Alderson, just arrived on the new railroad, tells of the end:

That year [1882, along the Tongue River] the buffalo were still so thick that Mrs. Lays had only to say: "Mr. Alderson, we're out of meat"; and he would go out and find a herd and kill a calf, all just as easily as a man would butcher a yearling steer in his own pasture.

Yet when I came out, one year later, there was nothing left of those great bison herds, which had covered the continent, but carcasses. I saw them on my first drive out to the ranch, and they were lying thick all over the flat above our house, in all stages of decay. . . .

The summer after I came out [1884] Mr. Alderson killed the last buffalo ever seen in our part of Montana. A man staying with us was out fishing when he saw this lonesome old bull wandering over the hills and gullies above our house—the first live buffalo seen in many months. He came home and reported it, saying: "Walt, why don't you go get him?" And next morning, Mr. Alderson did go get him.

I lean on the railing, watching sun glint silver off the backs of brown logs, floating by like waves of buffalo through prairie grass, carcasses of trees as far up and down the river as I can see. The logging roads push in, like the railroads, the big, old living things are taken out and sold, land ownership changes hands, and a new economy is established, geared to large volume and export, to centralized control and centralized, short-term profit. We knew the buffalo were going; and here, even ministers and timber companies say there is little time left in Sarawak. But is that simply bad forestry? Will they even miss the trees? Or, like the buffalo, do the trees stand in the way of tree ranches, plantations, roads, and retail economies that will complete the transfer of power over the land, from native residents to absentee Malays and Chinese? That is the transfer of power over the land that Rajah Charles had feared in 1915: "strangers and speculators . . . will, in their turn, become masters and owners whilst you yourselves, you people of the soil, will be thrown aside and become nothing but coolies and outcasts of the island."

Then I remember that D'Arcy McNickle had said native land rights were the central issue in the relations of whites to Native Americans. He knew the subject well—a Salish Indian from Montana, anthropologist, Bureau of Indian Affairs official in Washington, professor, founder of the Newberry Library in Chicago. He said that a tribal, communal tradition of land rights simply couldn't understand either private landownership or the kind of exploitation that seemed to accompany it. In the east, the Cherokees by 1881 had learned and dissented: "The land itself is not a chattel," they argued in court. McNickle in 1959 said:

Even today, when Indian tribes may go into court and sue the United States for inadequate compensation or no compensation for lands taken from them, they still are dealing in alien concepts. One cannot grow a tree on a pile of money, or cause water to gush from it; one can only spend it, and then one is homeless.

The slaughter of the buffalo is often mentioned as one of the worst environmental disasters in the history of the earth—so many, gone so fast. And in their place the cattle quickly destroyed the buffalo grass, which was adapted to different grazing patterns. By 1886 the Montana plains were badly overgrazed. The entire ecology of the high plains changed in fifteen years. And here, in Southeast Asia, the disappearance of the tropical rainforest may be an equivalent disaster: since 1960, in the Philippines, Thailand, Indonesia, New Guinea, Borneo—gone. And if the buffalo were killed only for their tongues, what of the trees going to disposable plywood in Japan? How many brown giants must be killed before . . . or does someone want that change, in order to take control?

The more I think about it, the more depressing the analogy becomes, and the closer the Borneans and Native Americans seem. When was the land bridge to America? And when did the Bornean ancestors descend from the steppes of Asia? Are the Penan a reincarnation of the spirit of Crazy Horse?

The boats lands with a thunk at the dock at Long Lama, the upper end of the Marudi express run. The throbbing engine shuts down. Silence. We walk up steep notched logs to the top of a rise. One dusty street runs along the bank, open to the river, while along the far side are three cafes and seven or eight Chinese stores with trade goods. Richard takes us to the end of the street and points out the modern buildings and playing fields where he and Harrison Ngau and all the others in the middle Baram went to high school. In the twelve-hour journey from Marudi to Long San, this street is the only chance to buy a bowl of rice and fried vegetables. The express crowd is filling the cafes; dogs and children run in the street. We are introduced to a Chinese shopkeeper, a friend of Richard's, who at the rear of his tiny store has an office with a radio telephone (you can call Marudi for a few dollars) where he writes and prints a local paper. In the crammed shelves of his store, maybe ten feet wide by twenty feet long and packed to the ceiling, I find a little round box, made of handwoven rattan, with a dozen coasters inside. Coasters. For martinis and highballs on your coffee table. The box is covered with dust. What in the world are these coasters doing in Long Lama? "We've had them for five years at least," he explains. "A government person said that's what tourists want, and some women wove them from vines." Do you get many tourists here? "One a month, maybe." We agree on half price: three dollars. I buy. Jim and Lois Welch will love them. Richard thinks it's a fair deal all around.

After a lunch of fried noodles with tiny dried shrimp which you eat whole, crunchy but tasty, Juliette and I wait for our express, sitting on a bench in the shade, overlooking the river and the busy dock. There is a noise in the tree above us; locals scatter; a coconut hits the concrete bench a foot away from Juliette, bounces three feet up and rolls off into the bushes without breaking. We go down to the dock.

Above Long Pila, the Baram narrows. We are leaving the lowlands. Hills come down to the water's edge, pinching the river into brown swirls, and for the first time we see virgin jungle. Tom Harrisson in 1932 noted that little virgin forest could be seen by boat all the way up the Baram to the Tutoh, since natives cut the riverbanks first. Now the commercial as well as native logging has reached Long San, and easily approachable river frontage is always cut. But steep hills without landing places are here untouched, and as the boat winds up through a narrow passage, immense trees rise on both sides. We begin to feel the magnificent presence of that forest, trailing vines and lianas and flowers into the water. The growth presses on our senses, and we look forward eagerly to entering that world, somewhere above Long San. When it is not so steep, hillside rice fields — shifting cultivation — appear. This natural beauty, however, brings no one else up from the wrestling on the TV below.

"Slash and burn" or "shifting" agriculture is a hot nondebate. That is, scholars have decided that traditional slash and burn practices (called shifting cultivation, swidden agriculture) do not destroy forests. Cutting primary forest is a great deal of work. Natives cut three or four fields, clear and burn, and plant them in succession (mainly rice), letting the others lie fallow. It is not in their interests to allow erosion, to destroy fertility, or to cut too much. If population does not increase, after a certain number of fields are cut, no more primary jungle will be cleared. Subject finished.

Timber interests, on the other hand, point out correctly that shifting cultivation involves clear-cuts and wastes trees, while lumber companies practice selective, lucrative logging, and that large areas can be completely denuded by natives (near large population centers, this is true). James Wong has said many times in print and in person that without doubt shifting cultivation, not logging, is destroying Sarawak's forests. One wonders, however, why through the last thousand years the forests have remained, while in the last fifteen years they have disappeared. Busy natives.

Aware of this nondebate of opposing certainties, I looked carefully (usually the only person above deck) at the Baram River, over several trips, from its mouth to within a few miles of the top of navigation for a longboat. I saw thousands of acres of swidden land, either fallow or bright with rice crops, often on very steep hillsides, but I saw only three swidden clearings with significant erosion. The erosion totaled about ten acres. This equaled, approximately, the single natural landslide on the steep ridge near Long Anap, or any single junction of two logging roads on a steep ridge anywhere. To make it clear: judging by what can be seen from the river, all the swidden erosion in the Baram is equal to about one mile of logging road erosion. The Baram was always muddy after a rain, say the people of Long San, but it used to clear up in a few days. Now it never clears.

We continue up the winding, narrow river, past the lone cabins at rice paddies and past settlements where children stand at the top of the bank, brown and waving, or run naked up and down the logs moored in the eddy while mothers and older sisters, in bright sarongs, kneel with plastic pails and suds, rubbing their wash on the wood. We stop wherever someone waves a white handkerchief, the giant boat nosing up to the one tied log or into the mud bank as someone hops on, or more often, hops off laden with bags and boxes from Marudi, greeted by a family of six children and an old woman and a tail-wagging dog. As we back off and roar upstream, they disappear into the jungle.

We pass Uma Bawang, home of Jok Jau, who filed suit against the timber company for destroying cultivated land, who hosted the blockade and election celebration in October 1990, and who at that moment is landing in Yokohama with Jewin Lihan to testify against logging, and then Long Kesseh, the small Kayan longhouse where Harrison Ngau grew up. Two young boys cross the river in front of us in a ten-foot dugout, one paddling like mad with his hands, a wooden shingle held in each, and the other bailing just as fast. They laugh as the huge express toots at them.

Finally we reach Kelawit camp and Long Na'ah, end of commercial boat travel up the Baram, and the beginning of Samling territory. Samling's Kelawit timber camp is a hundred acres of shacks, bare mud, dust, and timber. Front-end loaders roar past with logs, dunk them in the river and leave them for cowboys to corral as they float away. As the boat noses into the bank, we are surrounded by noise, dust, huge engines, cranes, rafts of logs being lashed together. This is where the middle Baram timber comes out. We get off the boat. A company pickup provides all who wish it a cheap overland lift to Long San, saving hours of travel and if river conditions are not right, days, for above here the rapids begin. We drive for an hour along ridge tops and through dark ravines; beautiful, heavily forested mountains rise steeply around us. Before dark we drop off a ridge and descend to the river at Long San—the improbable green soccer field between the neat mission school and the church with its beautiful Kenyah designs, the government school compound with its twenty or so buildings to board students from up and down the Baram, the 150-yard longhouse and most improbable of all, the Tattan Hotel, a house on stilts with three rooms to rent, where Juliette and I discover we can retreat from communal life with a private room, a bed, and a serviceable public toilet and shower for $6 a night. Many fascinating and exhausting longhouses lie behind, and ahead. We retreat.

We have hardly unpacked when I hear a strange, tiny melody. I creep out the door and down the hall to the family sitting room. On the wall is a Fuji clock, with three AAA batteries and a computer chip, playing "East is east and West is west and the wrong one I have chose. . . ." I go back to the slat room, and look out at the pigpen and banana trees. Juliette sings, "And I'm all yours in buttons and bows."

Above Long San, primary forest begins, unroaded and back from the rivers, untouched. We have finally made it up the Baram. The next day, Richard, healthy, handsome, articulate, and recommended by Baya Timber, needs only one hour at headquarters to be hired on at Samling, filling in for sick help until a permanent opening should occur. Over the next few days we are able to enter the forest above Long San, in the Lambir concession, and inspect trees about to be cut.

We camp for two days under one exemplary tree, a large meranti, as the cutting moves closer.

Because the feller (sawyer) for that area was killed by a falling tree Saturday, November 17, 1990, Richard (on loan from Samling), his friend, and another boy from Long San took over the dead boy's work, and sneaked us in to our campsite under the tree. We had known that our meranti would be cut within a week. We had not known that Richard would cut it, or that we would be there to watch. We were beginning to learn what "upriver" means.

CHAPTER 5 🦎
TREE, FOREST, LOG

Monday, November 19, 1990, Lambir Camp

We are just above Long San, at Lambir, the last timber camp up the Baram. The logging ends a few miles away. Beyond that, a sliver of road inches upward, a long red track through virgin forest, to the point where a bulldozer sits. It has been there since August. One vine has already laced itself through a tread, and in two years the yellow hulk, if left immobile, will look hardly more improbable than the hanging orchids and thirty-foot high ferns which surround it.

We know the story of why the dozer stopped, for we have met the three Penan who stopped it. In August they found a surveyor with his equipment in the virgin forest near their settlement on the Salaan River. Their account:

"What are you doing here?"

"Oh, just looking around."

"Who do you work for?"

"Nobody, just looking."

"Why does all your equipment have a Samling emblem on the side?"

He shrugs.

"You'd better leave, there is no logging agreement here."

"I am not logging."

"You are surveying for a road."

"I have nothing to do with the road. I am just surveying."

"But when you are finished, the road will come."

"I have no connection with that. If tractors come, you complain to them."

"Well, we should tell you that it's not smart to be walking around the forest like this with no timber agreement. You could be mistaken for a wild boar."

As always in the woods, they were carrying their blowpipes. I have sat on the deck of a Penan hut watching them plug gourds at twenty yards: ffffwht-thunk, ffffwht-thunk. Say it as fast as you can, with no pause. The darts went through the gourds in a pretty tight pattern, about four inches square. They also put them into wooden posts at the same distance. They had to be broken off. They tip the darts with distilled poison from the Ipoh tree; a deer dies in about an hour. If you picture these fellows, built like welterweight champions, naked to the waist, parangs, six-foot blowpipes with spears (to finish off the pig), a few ornaments or feathers, bright burning eyes, and sweet smiles, you may appreciate how the surveyor felt. He left, and the tractor at the end of the road up from Lambir has not moved since.

The Lambir camp works a small concession about three miles above Long San. The logs are dumped into the Baram at this bulldozed, mud ramp. Along the riverbank are three barracks, an office shed for the Chinese grader and the native yard workers, and ten large fuel storage tanks. Up a steep mud bank are the administration buildings; behind them, on a cleared flat, is a semicircle of equipment sheds and at the top of the hill is a small Chinese temple, a three-sided, bright red wooden shack with food offerings, incense, and a little Buddha looking out over the Baram, down to the new bridge pilings. Back behind the temple is a huge roof covering nothing: the site of a lumber mill which was never built. Instead, the logs go out whole. Some say Lambir ran out of money, some say inefficiency has caused delays, some say a promise to build a mill can help secure a timber concession. The Sarawak Plywood Mill downstream has never turned out a single sheet, but its existence enabled someone to cut and export whole logs.

Around the unfinished Lambir shed are piles of logs, big hardwood logs two to six feet in diameter and thirty feet long, which once awaited the grand opening of the mill. They are already hollow at the center, rotted and useless, according to workers at the camp. I count 430 logs and quit. That would be about 300 trees, or 200 acres of the tribal lands of Long Selatong Ulu, simply cut and left to rot.

Everyone says Lambir will sell out to Samling after the bridge is finished in June. Samling's Long San bridge, visible to the Buddha a mile downstream, all five pilings in and four of them spanned by girders, will carry heavy equipment to the upper Baram. Like a famous bridge on the Tutoh which was burned in 1987 (for which burning a number of natives — the wrong ones — were jailed), this bridge is crucial to opening new territory. By June, a legion of bulldozers may join the one now rusting at road's end, and the Penan will have their hands full. Whether they fight alone, or with something like legal status and community backing, will be decided at Long Moh, when the upper Baram accepts or rejects the timber agreement.

We have come to Lambir camp, Richard and a Long San cousin and Juliette and I, to camp for two days beneath a mature meranti tree in the path of logging. The natives know well what they are losing; we hope to gain at least some impression of virgin rainforest a mile beyond logging in the Baram, Sarawak, North Borneo.

TREE

I have sat for an hour in still rainforest, not a breath of air, and have heard three huge limbs come crashing down from the upper story, each bringing an avalanche of orchids, ferns, and vines connected to other rotting limbs, and those connected to others. One limb, twenty feet long, never reached the ground, but hung to gather moisture and dirt and seeds, and support colonies of ants and other insects up near the surface of this zone in which I am immersed like a diver at the bottom of the sea. This is not my northern world of ice and air, the clean hard edge of cornice curving against blue sky. This is something else.

I am standing in rainforest in the middle Baram, facing one large tree. But we are deceived by the word "tree," a clean noun from the world of hard edges and ice. This is a gathering, a neighborhood, a mob of vegetation; in fact, I can hardly see the tree.

Picture a stage set. Center rear is a large tree trunk rising through the top curtains, the lowest limbs just visible and spreading out of sight, into the wings. Around the trunk are twined, crisscrossed, running every which way, ten or fifteen ropes of different sizes and shades of brown and gray, so that the trunk itself is hidden. Five or six more ropes hang from the limbs, all the way out to the wings, some of them hawsers the size of your leg. The entire mass has been spray painted here and there with splotches of dull green, bright green, and russet, where patches of lichens and moss have spread. Colonies of flowers are interwoven with the clinging and hanging vines. The ground is remarkably clear, almost like a New England woodland. This is an environmental theater: the humidity is an invisible mist, although the air is pleasantly warm and fresh in this deep shade. The smell is life itself.

Do not ask the plot.

Let us approach. What I am calling a tree is a tree-trunk-vegetable mass, five feet across at chest height. At arm's length, I see first a thick, reddish trunk entwined with six or seven large vines and many small ones bearing shiny green leaves. Two other large vines drop straight from the upper branches to the ground beside me.

I step closer, nose to lichened vine. It seems that I have not seen the trunk at all. From left to right I count fifteen vine stems finger width to four inches thick, entwining the tree (which I assume exists). Between these easily noticed vines, within one inch I count four smaller vines, vein or thread size, all with many filigrees, lateral roots thin as the thinnest fishing line, the filigrees themselves becoming a hairy pulp. Most of the vines have differing leaves; how many species are present in this five-foot span I cannot begin to tell. Various insects are working the pulp. It is a green and brown mess.

Like a child with the impulse to touch the unknown, I reach out and push my index finger straight at the tree. It sinks in.

There is no line where tree meets air. My forefinger goes in two inches and more, and hits nothing hard. The bark is not encrusted with growth but rather permeated by it, expanded into a soft mass of life, a zone of vines, lichens, soil, flowers, parasites, and epiphytes which live on the tree without harming it. Within this zone, flakes of bark appear all the way to the outside surface. Yet even at the "surface" there is no line; there is no surface. Vines smaller than my pencil tip disappear into air so thick and humid that the naked eye cannot find the edge of growth. This is not a crust on a tree; this is a zone that defines the rainforest, where the substance "tree" dissolves into ambiguity or omniguity, for most growth here is marginal and codependent. At first, this zone looks a mess: neither air nor not-air (several middle-story hanging plants draw all moisture from the air instead of from soil or rain), neither tree nor not-tree (in the Amazon many familiar English shrubs are present as trees, and the biomass of a tree may be exceeded by that of its epiphytes and parasites, which may also outlive it), a zone where growth and decay are so simultaneous and interpenetrated that simple binary distinctions—life and death, for instance—break each other down in constant, mutual dependence.

I look at the right-hand edge of the "tree." Vines arm to finger width are wound and sandwiched together, the thickest outside, the thinnest inside melding into the reddish pulp of the trunk. The rough, gray skin of the big vine is covered with lichens, dark and light green. Where vines twist together, something grows in the gaps—here a space of half an inch is filled with molding debris and root mass supporting a colony of bulbed flowers, yellow, fingernail size, worked by ants; fallen twigs and leaves are caught in the flowers and the cracks between the vines. The three vines closest to the trunk, with strange protuberances and different lichens, are themselves crisscrossed by smaller roots and vines. I am Alice in wonderland; the world has shrunk to fit in front of my face. The gap between vines is a little theater, like one of those Russian Christmas tree ornaments: one side of the egg opens into an entire scene, framed by other vines, presenting at least a dozen plant forms. It goes on and on, this "tree," more verb than noun. I am glad I do not have a magnifying glass. I have the feeling of looking into a hand-held slide viewer, or a kaleidoscope: you peer into a tiny window, and a different cosmos opens up. I believe if I were a botanist, I could spend a lifetime studying the growth within the span of my arms.

The air and moisture are so rich that soil is built anywhere: some thumb-sized flowers are growing directly out of the moss-lichen-soil adhering to vertical vine stems, which themselves entwine the tree trunk. The dark green of a common, slender climbing vine gives color, along with occasional patches of moss an inch thick, and two or three other types of green plants beyond my ken, some looking like thumb-nail bananas or pineapples without their fruit.

This is not a buttressed tree, and at my feet the outcurving lines that should be roots may well be vines filled in with growth. Nothing is clear, since the ground is covered with decaying leaves the size of magazines. I push my pencil all the way into the soil with no resistance; the hard, infertile clay ground of the jungle is completely covered by life-fall and airy rot. The colors are those of a New England forest floor, with spring and fall superimposed: splotched brown and decaying leaves, green lichened bronze leaves, a yellow leaf that looks like an oak, and dull red elmlike leaves, all lie amid bright green plants and fresh seedlings and tiny tendrils juicy and delicate.

The light is filtered through so many layers and colors in the upper and middle canopy, that it arrives at the floor mottled and modeled, almost sepia. At two in the afternoon on a sunny day, I am surprised to see I am shooting ASA 400 film at a 30th, about one-third the light of a northern forest.

Not everything around me is vegetation. Every leaf I can see, large or small, high or low, has at least one bite taken out, and many are half eaten or decomposed on the vine. I never see what is eating them, though it is not for want of looking: after fifteen minutes crouched at this tree, in shorts and sandals, I am well known hereabouts. The heat-seeking leeches, thin as pencil leads, heads raised and waving, are heading for my feet from ten yards away, advancing across the leaves at a mile an hour. I have sprayed my feet and ankles, which helps. Some. I feel like a great white fortress of blood, under siege. The leeches do not hurt, and carry no disease. They begin as inch-long slivers that can enter a shoelace hole, and they swell to the size of eggs. The trick is to find them fast and pluck them fast, rolling them between the fingers as you toss — or else they will be fastened to your hand. They do what they do very well.

At my feet, an obstacle to the advancing leeches, is a bird's nest fern which has fallen, with bits of limb, from far above; it seems to be growing well on its side, its hairy mass — I try to lift it, it's over two hundred pounds — supporting a colony of orchids with banana-sized bulbs, and succulent plants which stored rainwater, up in the canopy. The edges of the mass are porous with roots and spiderwebs and decaying leaves. I look up at the canopy above, but cannot find whence the orchid came. I get out the binoculars; six or eight stories up, ferns and orchids are scattered like giant condor nests. Some clumps have fine roots dangling six feet down, like jellyfish tendrils, sucking moisture and nutrients from the air. Once in Sabah, at Poring, I spent a day on a walkway strung up in the canopy, an aerial path between treehouses. The shock was what lives so high up: colonies of ants, worms, lizards, all crawling about these clumps of growth and decay, eighty feet off the ground. I peer above me now — hanging from the canopy are the various lianas, or climbing vines, legumes, cucumbers, passionflowers, some of the hundred or so types of rattans. My neck is stiff; I lower the binoculars and come back to earth.

I look at the tree again, and notice where the flowers appear. On the trunk, any break or nick or ledge becomes a colony — the tiny wound in a vine stem in front of my nose has two fingertip flowers. Each bent elbow of twisting stem, and there are thousands between my feet and hand, collects mulch and moisture and supports new life. A six-inch gash slashes across one thick vine; inside the wound, various plants have sunk their roots. One pencil-thick vine rooted in the gash ascends straight up for thirty feet before attaching to another vine. All around the edges of the wound, like lips, green moss puckers.

The largest vine is a strangling fig four inches thick. The strangling fig begins as a seed high up on a branch, deposited by a bird. It sends a root down, fine as a hair, until it finds the ground. Then the vine thickens and branches, big as an arm, a leg, crisscrossing the trunk, weaving a basket around it. In a hundred and fifty years the tree dies of constriction and rots in place, but it does not fall. Held within the vine's embrace it can only collapse straight down on itself like a building demolished, leaving a latticework of leg-thick vine, a vertical column standing a hundred feet high. I have climbed up inside one, the trellised tube of hardwood vine six feet wide, rock hard and steady.

Two of the vines on this tree cross over to, or come from, neighbors, about fifty feet up. One strangling vine nearby has clearly overcome many smaller, short-lived trees, and loops through the forest for fifty yards, on the ground, up into the air, across to a trunk, on again, leaving a record, in vertical hardwood spirals, of the long-gone trunks it once entwined. To look at the living vine is to see the ghosts of trees. Here Thoreau's dictum that all seasons are present at once (he delighted to find the warm spring moss in snow) has come to a final equator: there is one season in which growth, maturation, decay, and death are indistinguishable. What is this mass of life-in-death, of death-in-life, this "tree?" Where does it end?

This tree is a *Shorea cortisi,* one of many types of meranti, a red hardwood, full grown to five feet in diameter and 130 feet high, but not a giant. This one is straight, however, and free of lower limbs.

The tree stands on a small flat near a clear, tumbling stream at an altitude around one thousand feet, in montane forest. There are many trees close but the canopy in this hill forest is broken, not nearly as dense as the lush lowland dipterocarp forests, which in Sarawak are gone. Standing here, I can count three major and thirty-five minor trees over about two acres of vision; the nearest large tree to the meranti is forty paces away. Richard and his cousin say there are three or four marketable trees nearby; they point out four smaller ones, not yet ready to be cut, which may be damaged when the big one falls.

Connecting vines above create havoc; you never know what will come down — flowers, plants, rotting limbs the size of elms from vine-linked trees fifty yards away, and the trees themselves.

The ground is uncluttered, more like our "woods" than a "jungle." "Impenetrable jungle" grows mainly along riverbanks, in clearings or cut forest, where increased sunlight favors more leafy vines and an entirely different ecology, including different tree species. Deeper in the virgin forest the ground is quite open, with occasional thickets of nasty shrubs or vines, usually where a giant tree has fallen or water gathers. The Tarzan films and a thousand silver screen machete sequences must have been shot in second growth. Jungle walking is usually a good deal more pleasant than we think, except for the leeches and the terrain, which in hill forest is constantly broken by ravines and ridges and vertical cut banks, all steep and slippery clay. Leave the machete; bring crampons.

To the casual observer, the mature trees on this flat would seem to be of four or five types, red or white, smooth or rough bark, but looking at the leaves through binoculars (they begin about fifty feet up), one has trouble finding two trees alike. In fact, in one acre of tropical forest you may well see twenty, thirty, sixty, eighty species before a single one repeats. This is in shocking contrast to temperate forests where mixed oak, maple, elm, beech, and birch would constitute a rich variation, and where in pine, fir, or spruce forest one species will comprise 85 percent of the trees. Around our home in Montana, ponderosa pine dominates areas the size of Sarawak. Great Britain is twice the size of the Malaysian peninsula and has 1,430 plant species, to Malaysia's 7,900. In a sixty-acre sample of Malaysian rainforest, 381 species of trees were counted, almost half occurring only once in the sixty acres. The great naturalist Alfred Russel Wallace, the friend and guest in Borneo of the first Rajah, James, in the 1850s, wrote of the rainforest:

> If the traveller notices a particular species and wishes to find more like it, he may often turn his eyes in vain in every direction. Trees of varied forms, dimensions and colors are around him, but he rarely sees any one of them repeated. Time after time he goes towards a tree which looks like the one he seeks, but a closer examination proves it to be distinct. He may at length, perhaps, meet with a second specimen half a mile off or may fail altogether. . . .

The commercial taxonomy for tropical timber — "mahagony," or "meranti" — is very crude, partly because it is advantageous to lump ten or fifteen similar woods under one lucrative label ("mahagony") and partly because forestry Ph.D. would have trouble telling what you're cutting.

In tropical forest, the diversity of tree species is matched by a diversity of other flora and fauna without parallel on earth. A four-square-mile section of rainforest contains up to 1,500 species of flowering plants, 750 species of trees, 125 of mammals, 400 of birds, 100 of reptiles, and 150 of butterflies. The forest is full of plant and insect species yet unknown to science, many possibly useful to man, marketable or medicinal. One of the most powerful arguments for cutting plantation or monoculture trees, and not the rainforest, is that we do not even know what is in our oldest, richest seedbank. Yet we turn it into plywood. This seems especially unfortunate when we have hardly begun to explore synthetic wood substitutes.

Everything in the forest is subject to the most intemperate variations: stick bugs ten inches long, grasshoppers the size of cigars, moths dull black above and electric blue beneath, orchids the size of a pinhead, flowering tree trunks, phosphorous toadstools you can read by at night. A fig tree in fruit can attract fifty species of birds.

When the great ice sheet descended over and over again like a white shade to darken half the world, it stopped just north of the Borneo forests. Life here has been evolving with a minimum of traumatic interruption for over 100 million years; 150 million is the common estimate. The tendency of evolution to vary species begins to explain, along with heat and moisture, the extraordinary variety as well as fecundity of this particular equatorial forest.

To put 150 million years in perspective: land was first colonized by amphibious frogs and lizards 300 million years ago, and 150 million years ago the land still belonged to the dinosaurs. That was not the same land we know, however, for the continents had just begun to drift apart. The continuous evolution of this forest is as old as the continents, as birds themselves, or flowering plants. Mammals evolved only 50 million years ago. North Borneo was under water 30 million years ago (the forest migrated back and forth from Kalimantan); the mountain spine of the highlands was uplifted 15 million years ago, and Mount Kinabalu, at 13,500 feet the highest peak between the Himalayas and

New Guinea, is only 2 million years old. In those 2 million years, the ice ages occurred. We like to think of geologic time in the superlative, but mountain ranges and ice ages are children to this forest.

Naturally, the forest of 150 million years ago—before flowers or birds—did not look like this one; the point is that the gene pool in Southeast Asia has evolved and diversified, probably with a minimum of sudden interruption, during the entire history of man, mammals, and continents. Most of life and earth as we know it has grown up with this forest, and with few others.

Now the Philippines has been cut, and Sabah; Thailand and mainland Malaysia forests are gone. The Burma forest is being sold illegally to idled Thai loggers by the Burmese army, which needs money for arms to fight the Karens, and by the Karens who need money to fight the Burmese to defend their homeland: the Burma forests. Sumatra and New Guinea are seriously threatened. The great Southeast Asia rainforest is almost gone.

The primary sensation, standing here in the remaining forest of which this tree is a part, is of immersion in an extraordinarily rich and various system. As I look at this trunk enmeshed in vines and moss, deep in the shade of other trees, no two the same, I have a sudden, chilling memory of a perfect opposite: the windswept plains and snowy mountains of Montana, the isolated grandeur which I call home. On the east side of the northern Rockies, a vista of a hundred miles might include four or five types of grasses, one type of tree (cottonwood), five horses, fifty cattle, and a ranch house. The gophers are down below, the coyote is hid. Prairie and sky, peaks in the distance. One pickup truck is raising dust, at sundown, on the road to the ranch.

Now this forest, here, is something else. It might change the way one thinks.

FOREST

I have never felt less threatened by insects. After my bath, I am sitting naked on a rock in the stream, forty yards from the tree. Insects are everywhere, doing their daily chores so busily that it is hard to resent their touch. Very few things in the forest are really after *you:* leeches and mosquitoes, some rare parasites, a few horrendous worms. The other 30,000 species couldn't care less. They may want something that's stuck to your greedy fingers or exuded by your overfed pores, but they don't want you.

Where my legs emerge from the water, twenty or thirty fleas are cleaning me. I have no idea what they eat but their long probiscus mouths don't penetrate. They just busybody about, sucking from my skin whatever salt or moisture or oil they want. They tickle, of course, but that problem is covered by the first chapter of any meditation manual.

Now four fleas are on the page of this pocket journal, crawling toward the head of the paragraph, their long probiscus mouths going up and down. What do they need? Wet ink, semicolons, refreshing metaphors? Alas, they quickly flit away, the little critics, three by air and one by sniffing up my pen and finger and wrist, as if to trace this rot to its source. They were not interested in me until I had bathed. Now *that* is an insult.

A dragonfly comes, quite small, with slender green body and the most amazing blue wings—not quite the electric blue of the black moth nor the deep blue of the river kingfisher but a luminous blue on translucent wings.

The skaters look familiar. They are skitting about the surface of the pool where I have bathed. There was one when I came; now there are five. I must have added some spice to the eddy line. The skaters seem to use the ripples as a natural camouflage: if I move my leg they work busily in disturbed water, and they are almost invisible, but if I let the pool calm, the dimples on the surface from each skater foot are clearly visible and they stay near the rocks. If I pass my hand slowly overhead, they run from the shape or shadow; perhaps something attacks them from above. The blue and green dragonfly is back with another, and they dance downstream. My favorite butterfly, the Tree Nymph which seems to live hereabouts, drifts back into view. A pair, as usual, each larger than an open hand, with white, almost transparent wings, they glide and fly in dipping, sweeping motions, always about to alight and never, in my sight, doing so.

When I look back down to the stream, I am shocked at the different shadows. The twelve-hour equatorial day passes quickly and visibly, partly because light and shade are here reversed. Instead of the shadow of a tree or a pole or a house measuring your day, here columns of light pass by as the sun moves over holes in the canopy. Our clothesline has an hour of direct sunlight, between one and two o'clock. Every ten minutes, when I look up the log and boulder-choked stream, something different is shadowed, something different is lit.

We can hear four or five kinds of birds, all in the barbitt family, and various types of cicadas which go off at different times. It is three hours until my favorite, the six o'clock "gin and bitters bug" which used to call the Rajah's agents to the bottle. There are no mosquitoes or sweat bees today, no leeches ever in the water, and the yellow swarming butterflies from the beaches do not haunt this stream. A one-and-a-half-inch fish nibbles my toes and a half-inch catfish is working up my thigh. I am wonderfully appreciated.

A lone bull moose at bay against wolves in white snow—that is Jack London's favorite image for the condition of nature. Man against man, man against beast, beast against beast. "The law of life is meat," he said. "Eat or be eaten." But the law of this jungle may not be the same as London's law of the arctic, which was also America's turn-of-the-century law of business. Here it seems that "Eat *and* be eaten" comes closer to the truth, a more sociable formula: mutually assured destruction, and—a big difference—mutually assured creation. Here, the law of competition, a "struggle for survival" conceived as a conflict between two individuals, seems hopelessly simplistic. Here everything is eating and being eaten in such bewildering fecundity, everything is decaying and growing so fast, that the cold tableau of an individual struggle, one moose isolated, silhouetted against the northern lights, seems like colossal egotism. All that emphasis on one thing alone. Who cares? One moose more or less, one less me or you.

That is why, try as I might, I cannot feel the rainforest as "me versus them," although in grizzly country I have relished that pit-of-the-stomach thrill. Me over here, bear over there. Will they meet? Instead of a drama of "me versus them," the rainforest seems to be It, or Us, meeting constantly, going up and coming down at once, a thousand living forms intertwined at every step. The system is so immensely present, so immanent, that deep inside you dissolve like the tree, your hard edges peeling off into soft layers of support; parasites and lovers at every pore.

WHETHER OR NOT the forest is a huge supportive community to the Penan I will probably never know, although a helpful interpreter might cause the words to come from their kind mouths. However, from conversations I am fairly certain that they have no empirical concept of the ecological system as system, even if they are experts in each of its parts. Just as the concept of nature grew strong in industrial countries that were busy wiping the thing out, the thought of the whole system has not been necessary in Sarawak because as a system it has never before been threatened. When ten acres are cleared for rice, a native can tell you what grows back in sunlight and what doesn't, and so in that instance distinguishes a clearing ecology from a forest ecology. But such a clearing is so minuscule that it, and all the others, cannot begin to threaten the surrounding forest itself, the gene pool, the entire self-generating web. Therefore the macro concepts — gene pools, as opposed to specific plant-animal relationships — are not in the native conceptual vocabulary. Natives upstream of the logging can imagine being paid very little for valuable logs, can imagine their favorite trees, fields, or forests being bulldozed without their consent, can imagine many trees cut down and hunting ruined, but they cannot imagine the problems of regeneration in a changed ecology, cannot imagine so much cutting that soil begins to change under so much light and so little detritus, that patterns of rainfall are altered, as in Thailand, where vast areas of deforestation have increased heat and curtailed rain, so that now the land yields one crop of rice, instead of two. Unfortunately, this effect was not figured into the original cost of Thai logs. The Penan cannot imagine the consequences of a clearing, or of a selectively logged tree plantation, the size of Sarawak. Never before have such categories been relevant.

Our grand new scale of technology has created, among Europeans and Americans, a commensurate scale of imagination. The Penan know the trees, and what lives in them. We know the forest; it is our invention. These new ideas come to Sarawak from outside, and they are not dropped by birds. They are dropped by visitors, by the likes of us — writers, biologists, scientists, foresters, liberals, environmentalists, Greenies, activists, adventurers — whose migrations, habits, eccentricities, writings, and droppings are becoming a significant part of rainforest ecology.

Grace's favorite hotel is the Oriental in Bangkok, just voted for the tenth consecutive year the best hotel in the world. In the spring of 1990, I had breakfast with Grace on the terrace. It was two and a half minutes from the time we took our seat until a waiter brought us a menu. Grace says that's good. She is from a well-to-do British family, and had just come back from Borneo with a handsome environmentalist I cannot name, and was raising money for a purpose I cannot name, and by the end of the week would run off with yet another withheld name to stop the Narmada dam in India. There seemed something profoundly right about her ideological fidelity and global promiscuity, at once drifting and professional, as if most of her species were on the make, dreaming of issue-affairs, jealous of competition, longing to possess some tribal tragedy and call it their own. This is unfair to every liberal do-gooder I have met; and yet, for every one, is just a little bit right.

The story of Chico Mendez, the murdered Brazilian rubber tapper, has been slow to reach the screen partly because three different environmental groups were fighting over the movie rights. The world tour of Penan in the fall of 1990 with Bruno Manser and Thom Henley of Endangered Peoples Project (of Canada, not the same as Society for Threatened Peoples, a Swiss co-sponsor) caused some falling out with SAM (of Penang and Marudi), whose Penan were being abducted — or did they elope? People who should have been consulted were not, and were surprised, etcetera. The Consumers Association of Penang (CAP) is perhaps the best activist organization in East Asia, parent to the Third World Network and, with Friends of the Earth, to SAM. CAP is extremely straight, dour as an old Marxist, suspicious of all first-worlders. Yet they break the Borneo forest news that Greenpeace and the monkey wrenchers need to hear. CAP is as odd a bedfellow for Earth First hippies as Garry Trudeau's Chinese comrade shacked up with Duke.

In addition to personal tiffs and ideological squabbles are more serious structural problems. In between several trips into Borneo, in October 1990 we flew from Bangkok up to Kunming, China, to attend a meeting of the International Society of Ethnobiology (which could not invite Bruno Manser because if they did, a certain rival environmental group might boycott). Darrel Pusey, a leading Amazon ecologist, explained the situation. An environmental group most easily solicits funds for a specific purpose: save the whales, the bears, the Amazon forest. When they approach corporate or governmental sources, as well as individuals, it is good to be the major savior of whales, and even better to be the only one. Another group springs up in Australia with a glossier brochure and bigger names on their masthead, and your office may be in deep arrears. I asked Darrel if there was some central agency or clearing house that coordinated activities (one California group is called Rainforest Action Network) and could keep one informed of all upcoming rainforest whatever—boycotts, rock concerts, passions, crucifixions, salvations. He said not a chance. The World Wildlife Fund and Friends of the Earth are leading players, but the team has no coach.

There is another deeper and sadder reason for what we call the fragmentation of the left, which is really the fragmentation of citizen activists or of those who on behalf of the many try to loosen the grasp of the few: activists are usually out of power, and frustrated, and so often what they do makes little difference. They work for decades to gain an inch, and then one corrupt government and two billionaires set them back a mile. Sometimes they take it out on each other and themselves. As a member of my academic department observed during a long afternoon meeting, bickering increases when nothing is at stake.

Those in power may hate each other, but when the advantages of cooperation are so dramatic, they often perform wonders of diplomacy. Without the pressures of necessity or the inducements of wealth, those out of power are more susceptible to ideological, or egotistical, schisms. We once gathered on the Berkeley campus to mount a demonstration for the hiring of more blacks in Oakland restaurants, but the demonstration never left the campus because the Leninists and the Trotskyites—totaling about five of the five hundred people present—could not agree on the route. We were certainly not much help to the blacks in Oakland, as the Oakland Black Panthers gleefully observed. The splinterings, jealousies, and occasional cross-purposes of activist groups are heartbreaking but only too intelligible, as any activist knows; you try to maintain the vigilance, the counterpressure, the energy year after year—and the moment the citizenry nods, someone has taken their land and rights. In the case of native rights, an active citizenry must often be created on top of tribal structures, after the rights have already been taken away. It is not easy work, out of little offices with two phones and no fringe benefits, no Jaycee banquets, no girls jumping out of the cake.

Even if the environmentalists of European background were to work together as they rush back into the niches their retreating empires have vacated, they would still carry cultural baggage cumbersome in the third world. There is good reason for the Third World Network's suspicion of first world work. The first problem is Romantic primitivism, which sprang from industrial Europe in the late eighteenth century and blossomed in the American West. European primitivists invented the "noble savage" and sought to save the sacred earth from the evils of human civilization. No "primitive" tribe, however, considers its civilization evil, and only colonialists thought those tribes were uncivilized. So a healthy fear of *industrial* civilization is unhealthily hidden by European desires for a "natural" and "uncivilized" life. It takes a very sophisticated Earth First activist to distinguish such escapist romanticism from an ecological yet prodevelopment stance. The differences, however, are intuitively obvious to third world ministers and natives who want intelligent, fair, and sustainable development, not "wilderness." Over and over, with certain Chinese friends in Kuching, I found myself taken for a "keep-the-quaint-natives-in-the-stone-age" romantic.

For instance, Grace was interested in the Penan Biosphere Preserve and in the creation of more parks in Sarawak, but had never thought about logging rights and methods appropriate to the longhouses. It is a good test case of first world versus third world environmental differences: a gigantic park in Sarawak might help save the rainforest, the ozone, and counteract global warming, while destroying longhouse life. Natives would either be moved out or doomed to a museum economy. The great strength of SAM is that they receive global research on "nature" (that European obsession) from Friends of the Earth, while their own native staff begins with an interest in a viable economy, long-range stability, and native land rights.

The greatest victory might not be a park, but Mitsubishi manufacturing (and the Sarawak government distributing with low interest loans) lightweight, portable winches, cables, saws, and outboards that enable natives to harvest whatever timber they wish from their own tribal lands. Remember the economics: with land rights secure, a native longhouse member could cut one average tree from tribal lands, yielding one log of five hoppus tons (worth $1,000 U.S. at Miri), and with a thirty horsepower outboard he and a friend could easily float three such logs down to Miri in a six-day trip. Allowing $100 depreciation on equipment, $200 food and lodging, and $700 for gas round trip, the three logs could be sold (to a Chinese sawmill) for a profit of $2,000, an income roughly equivalent to a year of family rubber tapping. How long would the rainforest last? Natives have been managing their own forest districts quite well for generations. How the money would be distributed, how often the community needed that much cash, and how much yield could be sustained from the longhouse lands—that would be up to each longhouse, and from observation I believe Sarawak natives are perfectly capable of making such decisions. Pessimists might not agree, but in the scenario of locally controlled exploitation, there is at least *incentive* for stewardship, the profits are local, and abuse, if it occurs, is self-determined.

The Japanese trading houses will point out that the tiny volume of trade thus generated would not create the market now present at the mouth of the Baram. That is to say, the first world is not interested in helping the third world extract on a sustained yield basis; the lowered volume might raise the cost of plywood a few cents a sheet.

Such discussions of modes of development and distribution of profit, however, rarely arise among first world "primitivists" looking for parks, wildlands, and a "traditional," "unspoiled," or "authentic" native life. That is, among Europeans looking for a lost past.

A second difference is that first world environmentalists tend to look out and see resource depletion and degradation—in the oceans, the mines, the forests. Lo and behold, such rapacious resource extraction occurs in the land of the poor. Oil is not found under golf courses; Japanese forests are carefully cut. Third worlders, on the other hand, look out and see first world consumption—in America of the 1990s, triple the per capita consumption of the 1950s—and they wonder where all that energy (their energy) is going, and why?

The ozone, global warming, shrinking forests—the landless poor of India, Nepal, Africa, and Brazil are not consuming three-fourths of the world's energy; we are. In August 1989, Indonesian President Suharto spoke for many in Southeast Asia when he said: "It is the process of industrialization which has taken place for so long in advanced countries together with the consumption pattern they have enjoyed for so long, which eventually constitute the greatest threat to the environment of mankind." The Sarawak chief minister, in 1990, said the crisis is in America, Europe, and Japan, not in Sarawak. Stay home and change your habits, they say; don't come to the forest and preach.

Finally, the third world still buys their machinery from the first world, and markets their products through first world systems. How much do we facilitate appropriate technology, small-scale or local development, local resource control? Not only do we consume, but in the process of consuming we *increasingly* encourage the concentration of wealth in the hands of the few: large technology projects, and monoculture plantations for export dollars to finance national debts, often bring rural subsistence farmers down to subsistence wage workers or worse, to the bottom: urban poor. These projects do, however, enter the ledger sheets of international trading companies, who gain nothing from successful, healthy, local barter economies. With each new round of GATT negotiations, some think the leading industrial countries are making a bid for even greater dominance over third world and small-scale economies. Unfortunately, capitalism does not always serve democratic ideals: individual freedom and self-determination of peoples.

As the political empires retreat, the first world economic empires seem to be advancing, and for the newly industrialized countries like Malaysia, it is not easy to employ the people in new factories without dancing to the tune of the banks and multinationals who fund the development.

So we come to Sarawak, Grace and Bruno and Thom Henley and liberal Japanese lawyers and American congressional committees and myself, on jet planes, on grants, junkets, or vacations, trailing clouds of absurdity. Natives know that; they also sense, and reward, kind intentions and compassion. Grace was very good humored about the absurdities. No anthropologist, she said, looking out over Bangkok, can really go native. I thought of James Willard Schultz marrying a Blackfeet bride in Montana in 1878 and writing *My Life as an Indian* without mentioning to readers that he went home to New York State each winter but did not tell his mother that he had an Indian bride. He was scared of losing his inheritance.

Grace said she loved the service at the Oriental—after all, her hippy days were over (not to mention a few years in the British Embassy) and she was doing what she could for those poor Penan. So we—an exotic species—come to these people and trees, and they to us, through looping lianas of information, transportation, and consciousness that are recently evolved, and that are changing the landscape. They know the trees; we know the forest.

LOG

Let's get a running start on Borneo—not the Borneo of escapist fantasy, but Borneo now. Why leave Bangkok? You can have an evening at the Maharaj restaurant on a deck over the river, live sitar music, spicy hot coconut milk soup, stuffed prawns, cashews fried with green peppers, rice, all for less than $5 including the soda and ice and a pint of Mekong whiskey. A town you could live in, though stinking and polluted, a Newly Industrialized Town with 20,000 Japanese and a strip of new factories stretching for forty miles to the east: Toshiba, Honda, Kawasaki, Suzuki . . . one American, one Canadian firm. An average income of less than $100 per capita and the best hotel in the world, which we couldn't afford. Boring clientele, however. I sat in the lobby of the Oriental thinking how expense accounts have changed the nature of the aristocracy.

We prefer the Eastern and Oriental Hotel in Penang, the understated elegance of the black cannons flanking the swimming pool. They remind one of Captain Light, the British founder of Penang, who hove to in his ship just offshore, rammed gold coins down the muzzle of his black cannon, and fired them into the jungle so his crew would get right to work clearing the forests—setting a pattern for Malaysian logging to this day. Or Norma's guest house in Penang, closer to the bone; dinners with Norma, an Indonesian-Malay Muslim, and Warrick, the twenty-fourth direct descendant of Edward the First, a retired Australian engineer and Plantagenet shopping for a sloop; the "Ancient Marina" Norma called him—yes, in her British-Malay high school she had read Coleridge, and *She Stoops to Conquer,* and *A Tale of Two Cities.* We were eating satay curry, okra, stuffed chicken legs, rice, and beer.

Then came our first trip to Borneo, out there beyond a service economy, we thought, where people are strange and real things happen. The first object I noticed in the airport in Kuching, in the window of a bookstore, was *WordPerfect 5: A Desktop User's Companion,* and on

TV, European soccer highlights, the jerseys Panasonic, Sharp, Phillips, TDK. The world comes in many forms: our sentimental candidate for favorite absurd hotel in Southeast Asia is the Damai Beach Holiday Inn on the coast near Kuching, $100 a night, palms and in-pool bar; in July 1990 at dusk, a small fishing junk appeared 200 yards off the beach, just beyond the surf. Two people swam to shore, cold and hungry. There were forty Vietnamese boat people in the junk. This was their first landfall; they had been in the South China Sea for eight days, and had run out of water. They couldn't straighten their legs. The hotel launch brought them in to the beach, gave first aid, sent five to the hospital and put the rest up in a poolside suite. The manager, Mr. Shamsir Askor, saw that they were showered, clothed, and fed the Malaysian buffet: barbequed prawns, fish, crab, curried meat, fried duck in plum sauce, satay chicken, salad, loksa, rice.

An Iban hotel worker is telling us all this on the bus back from the beach (we have seen a newspaper clipping and have asked). What's happening in his home? It's far upriver, he says calmly, the logging is a disaster, his parents will probably come downriver soon. He laughs. They will be boat people too, after the logging. Will they be fed the buffet? we ask. Not likely, he replies.

At Sibu in the Mandarin Hotel we watch *The Flamingo Kid* (poor Brooklyn boy falls in with swells at a 1950s Long Island resort), and on our first express boat, Sibu-Kuching, we pass twelve timber ships in the Rajang River registered in Indonesia, Panama, Singapore, Taiwan, China; the man seated in front of us, with broad back and broad waist, looks Native American and is dressed in cowboy boots, Levis, and Black Watch plaid wool shirt, but he and his wife and six children are speaking Iban says the policeman sitting nearby who offers a can of "Fizzi Rut Bir." More ships, women in bright sarongs doing wash on the docks while children jump in and out of the river, the sun burning through mists to shine off Nipah Palms in the Delta. Out in the bay, a marlin jumps three times and the rooster caged beneath my feet crows. At the blowpipe competition at the Kuching Fair I speak to a young native in coat and tie, an engineer for a government company; he is from the upper Baram where we are going; his parents and brothers and sisters all live there; no, the logging has not reached his longhouse yet. Thinking a yuppie will defend development, I ask if logging will help the economy when it arrives. "No, there is no advantage at all for

Niece Linda Bevis in lowland forest

Daughter Sarah Bevis in bamboo lowland forest

The Chinese storeboat at Long Terawan, Tutoh River, log raft

Samling's outlet at Tebanyi Timber Camp,
just below Long San in middle Baram

Kitchen of apartment at Long Anap

Juliette at Long Anap

Women in dugout paddling to rice fields, upper Baram

Long Moh, upper Baram

Long Moh, upper Baram, view from a kitchen

Long Moh, upper Baram, women in background weaving mats, with children, men gathering for meeting with Samling Timber

Crown of tree

Author and Linda Bevis on summit of Mt. Kinabalu, Sabah

those people. It's just terrible. It will destroy their life." An airplane flight to Miri up the length of the Sarawak coast, the primary forest all gone, and from the cab to the Baram River boat dock we count twenty-one ships in the South China Sea, lying to, waiting for barges with logs. Four to five thousand tons of logs per ship, the Chinese cab driver says. I ask what would happen to Miri tomorrow if the logging stopped. He laughs. "I dunno. They're getting rich now."

"And you?" I say. "It must be good for business."

"For a few only. They are cutting so fast."

"What about the children?"

"Nothing will be left. It is all very bad."

And finally up to Marudi, where in the SAM office all of Grace's favorite magazines are on the table: *Utusan Consumer* (CAP, Penang); *Asia Pacific Environment; World Wildlife Fund Reports* (Switzerland); *World Rainforest Movement* (Penang); *Alternative Law Forum* (Philippines); *International Society for Animal Rights Report* (Pennsylvania); *Social Alternatives* (Australia); *Panos* (Norway-Sweden); *Voices* (Hong Kong); *Marine Conservation News* (Washington, D.C.); *Probe* (Toronto); *Rainforest Action Network* (California); *Farm Chemicals International* (journal of the Asian NGO Coalition, Sri Lanka: "We appreciate the support provided by Agro-Action, Federal Republic of Germany"); and the *Peace News Nottingham,* which is advertising a job vacancy: "PN aims to be an equal opportunity employer, but regrets that there is currently no wheelchair access."

And in SAM's guest book, names, comments, professions from countries around the world, hippies, lawyers, "a concerned youth," government representatives, international environmentalists.

We come, from wherever we have come from, little high energy particles leaving little tracks, to Marudi, to find that we are preceded by ourselves. In my case, camping under a tree in the Lambir concession above Long San, I have come to realize that I needed to see primary rainforest, to meet people living a tribal life, and to observe what happens when a country like Japan or America comes to a poor and remote corner of this earth, coveting its wood. I want to write of another frontier closing. I want to see what we are doing to them, again. There is no sense pretending we are other than we are, or that Richard has the same consciousness; and yet there is common ground. There may be something plain wrong about a quick profit at the expense of your children, something plain right about caring for stability, for freedom, for the earth.

WE BROKE CAMP and left the forest and the tree to the loggers, thinking we would never be back. But things change. Just a few days later we found that we, not someone else in Lambir, would cut the tree we had come to know so well. Richard's cousin is a choke setter (he ties cables around felled trunks, so they can be skidded out to the road) for a tree feller at Lambir. The cousin sneaked us in to camp beneath the tree. That was good fortune, but by sudden misfortune events took a different turn. The feller, an Iban, was killed Saturday, November 17, by a falling tree. That was the first death at Lambir, and that same day another boy was sent downriver unconscious. Richard's cousin was moved up to temporary feller, since he had been working with the Iban in the same area, and Richard was able to hire on as his choke setter. Thus the four of us, with heavy hearts, climb into the back of a Toyota pickup to cut, among others, our tree.

We may never know what has brought us together, what vines loop through our little group standing on a flat above the Baram River; nevertheless we are gathered around the large meranti, which is tied to the forest itself. Cousin Johnny, who has never left Long San and speaks no English; Richard, Penan from Long Anap and Long Bangan who speaks so well; Juliette, the American professor of dance; and myself, the American professor of English. In spite of different origins, we trust each other. At our feet, the huge bird's nest fern still blooms with orchids; the leeches welcome us again; the huge trunk swathed in vines and flowers rises far above, into the canopy. Vines trail from upper limbs to the ground. A cicada whirrs above to our left; another honks behind. I pull my fingers down the trunk; flakes of wood and soil and moss and flowers cascade to the ground. I don't mind the tree dying; I really don't. The rainforest has taught me that. I mind the rainforest dying.

Richard glances at me. I nod, he looks away. He reaches down, takes the rope in one hand and pulls. The Stihl saw coughs. Instantly, cicadas cease. Unseen wings overhead beat to the west. He pulls again; it starts. I have a sudden, vivid memory of mahogany-red boards on a truck, the lumber clean and straight and sharp, beautiful in its way, hard-edged, like ice against blue sky. And I remember, from thirty years ago when I worked in a mill in Oregon, the fresh smell of ply-wood sheets off the cutter, right out of the pond. It was 1960; most of the workers had come West as Okies in the thirties, when the world was first united, in depression, just before Japan took Manchuria and Hitler Poland, before the first synthetic molecule was synthesized at DuPont, when America consumed one third the energy per person, and the world had half today's population, just before I was born. One of the workers, Dudley Culpepper, told Roger Smith and myself of shooting his year's limit of deer without leaving his privy. When forests still covered Thailand and Nepal, and in Borneo the upper Baram had never seen a white man, though they knew of the third Rajah Brooke, and Richard's Penan grandfather deep inside the Tutoh had no need to know of Marudi much less of an outside world, though he had seen Chinese jars, and when Harrisson's Oxford undergraduate expedition was exploring unknown Bornean peaks and pinning unknown butter-flies and drinking too much borak (a potent, smelly distillation of rice) in the lower longhouses under smoked heads hung from the rafters. All these lines of time and circumstance—do-gooders, millworkers, skyscrapers, the modern itself ("Light, noise, speed—you can never get enough of these" said Marinetti in London in 1909)—seem to con-verge on this little group of people from opposite edges of the earth, and on this tree, which stands hardly changed by the last half century, its strangling fig thicker, generations of insects, mammals, birds come and gone in its ample habitat, but for all that the last fifty years, on top of a hundred and fifty years of individual growth and a hundred and fifty million years of genetic growth, seem not to have mattered much, until now, when this tree, and this forest, like the Penan, will suddenly meet my world. Richard steps forward, gunning the motor. The chain rattles into a blur and blue smoke fills the air.

First, he will cut the vines.

PART II

BEAUTY AND MITSUBISHI

CHAPTER 6 ❧
SUNRISE IN JAPAN

The tree Richard felled November 19 in virgin forest above Long San produced three logs, and they were skidded out by tractor and cable, debarked at roadside, trucked by Mercedes to the Baram River at Lambir camp, graded at ten hoppus tons total, tied up in a raft of eighteen logs and nudged by two cowboys through the rapids down to Long Lama. They were joined to the big raft of a Chinese tug contracted to Samling, pulled to Miri, sold for $2,000, and shipped by a Japanese agent to an independent plywood mill just east of Yokohama. There our meranti tree became plywood. Most of the plywood shipments that month went to the construction industry, particularly to a company working on the new Shinjuku office building in Tokyo, and most of their plywood was used for concrete forms, the panels that hold wet concrete in place until it dries.

IN NORTHERN JAPAN, on upper Honshu, in Iwate prefecture, the streams run fast and clear. Fishermen with long poles and fixed lines stand in the cold water at dawn, angling for *ayu* and *yamame* — "mountain woman" trout. Around them, mists rise through cedars on steep hillsides, the dark greens and white mists and solitary still figures standing vivid as a landscape painting on an ancient scroll. That evening after a hot-springs bath, in the inn, you sit cross-legged on the spotless floor, dressed in the identical blue bathrobes of the hotel, around a low table, drinking sake, eating fish, telling tales with the other guests, all men, all boisterous and laughing and generous. They will not let you pay.

Outside, on darkening slopes behind the little country inn, the cedars are planted in rows. The small and well-spaced clear-cuts are carpeted with green ground cover. There are no erosion gullies, and the runoff is clean. Not a scrap of wasted wood is left on the ground. Japan takes care of its forests.

But the Japanese do not export wood from their forests or their forestry practices, and they import over 70 percent of the wood they use. Unfortunately, they do not treat other forests as they treat their own.

IT IS DIFFICULT for an outsider to write about Japan. What country has risen so far from ashes so quickly, on strength of character alone? For one who has lived there a year or more, the ties of friendship, generosity, kindness, and vitality are so strong—and yet. The Japanese are behaving no worse than my countrymen—and yet. The air is crackling with sudden, aimless, and dangerous charges against Japan—and yet.

The "and yets" come down to this: Many Japanese are painfully insecure, shy, and sensitive to what outsiders might think of them, and having little tradition of public dissent (though privately self-critical), they are especially sensitive to what others might say. An article on the front page of the *Japan Times* in 1990 began: " 'We must give more aid,' a Foreign Ministry spokesman said yesterday, 'or other countries will criticize us.' " One is tempted to believe that 250 years of isolation (roughly, 1600–1850) have left a mark. An island culture so strong and homogenous, yet Japan stands uncertainly in relation to others.

To many Americans, conflict is an integral part of public life, and criticism is part of friendship; to many Japanese, criticism is closer to betrayal. This goes beyond the usual and universal rule, that we can criticize ourselves but that outsiders had better watch out. Even within Japan, there is little sympathy for a "loyal opposition." Japanese friends who concur that they have no tradition of loyal opposition have often repeated to me with a smile, "Ah, you agree to disagree." They are right, of course, that our disagreements, if healthy, occur within a larger context of shared culture. It is no accident, however, that the wording they prefer gives the main verb to communal agreement, and thus they subordinate dissent. That is perhaps as close as they wish to approach our respect for fighting it out, a fair field with no advantage to either side, as Milton conceived the quest for truth, in a pamphlet against censorship.

Censorship is so pervasive in Japan that it hardly exists, for it need not be imposed from above. It begins, perhaps, with exquisite politeness, witholding reactions until you are sure of the other person's wants and needs; or at home, with holding your tongue in crowded family quarters. In a wonderful scene of *The Doctor's Wife,* a contemporary novel by Sawako Ariyoshi set in the nineteenth century, the newlyweds—bride hated by her stepmother—go to bed two feet and one rice paper screen away from the jealous mother. Careful behavior extends to silent subway cars of three hundred packed people minding their own business; Tokyo moves about three million people a day in and out of a city with no shouting, no excitement, no energy lost to friction. Working together—or better, being a member of a group prior to being an individual—*is* censorship as we understand it, an infringement on a thousand "individual rights," just as in the United States the exercise of individual rights necessarily infringes, constantly, daily, on society's right to peace, harmony, stability. If a human being is, as Plato thought, an animal that lives in a polis, or city-state, then America is an experiment in the animal's rights, and Japan, in the claims of the polis.

Of course, information can also be controlled from the top. The invasion, occupation, and colonization of Korea from 1910 to 1945 (involving slave labor shipped back to Japan to work in the Mitsubishi steelyards in Hiroshima, for instance, and to die there under our bomb), merited one paragraph in the official and mandatory high school textbook of 1990. The government censors struck the word "invasion" from the original manuscript submitted by the textbook authors.

This is not ancient history. In 1990, over half a million descendants of those Koreans still could not be Japanese citizens, and only in the last few years have fingerprintings of Koreans and job discrimination begun to decrease. In May 1990, when South Korean President Roh requested an apology from Japan for its cruel as well as colonial occupation of Korea, the ruling Liberal Democratic Party, after months of handwringing, finally authorized the Emperor to say that he felt the "deepest regret" for sufferings "brought about by my country" during that "unfortunate period."

The Economist, accepting the Japanese government's translation of the speech, called this a "handsome" apology. Yet the Chinese character *tsuseki,* translated by the government as "regret," was so obscure that not a single professor I knew had heard of it. They had to look it up. According to Minoru Hirano of the *Japan Times,* "This word was unfamiliar to the Japanese. . . . the word was also unfamiliar to the South Koreans. They doubted that the word meant an apology at all." A news columnist said *tsuseki* expressed a regret such as a batter might feel after letting a good pitch go by without swinging.

Even that hanging curve of an apology, however, rankled many Japanese. According to Takako Doi, chairperson of the Socialist Party, many in Japan felt that Korea's request for an apology was an unwarranted interference in Japan's internal affairs. Koreans once again contrasted Japan's reticence to Germany's Willy Brandt kneeling in tears in the Warsaw ghetto, or the German president observing the fiftieth anniversary of the start of World War II by saying the past must not be forgotten. A Japanese diplomat in Seoul said, "Other Asians are going to wonder how, if it cannot get things right with its closest neighbor, Japan will ever manage to lead the region as a whole." Finally, in the spring of 1991, Prime Minister Toshiki Kaifu returned from a trip to Southeast Asia during which he was repeatedly told that Japan, to be a leader in Asia, must face up to its past, and he ordered the Education Ministry to rewrite the chapter on Korea. Note, however, that the effective pressure for change came from outside Japan.

The day after the Emperor's apology I had lunch with two university juniors in Tokyo. She knew all about the Japanese occupation of Korea; he, a law student, was amazed. Japan, he said, never invaded Korea. If she had not enlightened him, he would never have believed me. She, it turns out, once had a teacher who went beyond the text. The law student, on the other hand, knew only what the Education Ministry had printed.

It made the news in 1991 that some Japanese scholars talked of the Japanese "rape of Nanking" in China, with much discussion of whether that was not altogether too severe a view. Letters to editors appeared saying "all circumstantial evidence is against the veracity of the massacre," etcetera. People had to reply quoting eyewitnesses, and naming brave Japanese veterans who have spoken out in public ("to tiny audiences") about their participation in the rapes and murders which continued for three months. The Japanese actions at Nanking were atrocious (and well remembered in China), and after fifty years one might expect awareness to move on to other, lesser crimes that are never discussed at all: such as internment camps in Borneo where civilians were murdered, sometimes for spite, sometimes for fun, and sometimes to ensure rice for the officers. Some Japanese know more of American imprisonment of Japanese Americans in Nevada during the war (Americans talk about it in shame) than they do of Japanese executions of civilians in their own camps.

Japanese actions abroad, from war crimes to strings on foreign aid to drift nets to whaling and logging, are seldom discussed in public in Japan, and even when mentioned, are heard by a very passive citizenry. In Tokyo, I presented my students at an English lunch table with articles on the Southeast Asian timber trade. They were visibly moved by Japan's role. "What can we do?" they asked. After several weeks of study and discussion, I suggested a university organization, Students for the Rainforest or whatever, and a campus newsletter to build support. They seemed enthusiastic. The next week, several were absent, and no one mentioned the rainforest. Asked why enthusiasm had waned over two weeks, they gave me answers that evolved from exams and insufficient time, through personal difficulties, to ruining one's chances of finding a job after graduation, especially with the big corporations such as Mitsubishi.

The legendary homogeneity, group consciousness, and conformity of Japan certainly exist—at least in postwar, rebuilding Japan (which may already be at the end of a phase). I met a college senior who spoke excellent English; it turned out that she had spent a high school year in California. As she left our English lunch table, she quietly asked her friend, in Japanese, not to tell others that she spoke good English or had lived abroad. It was all news to her best friend, who had known her since freshman year. Amazed, I questioned the five students and one professor remaining. Yes, it is possible her experience could be held against her—being different, setting herself apart—and yes, California might be held against her in matchmaking. What if she had different ideas? Would she make a good Japanese wife? That was June of 1990.

Within Japan, she might be perceived as *gaikoku kabure,* or "infected with foreignness." In a country that quotes adages such as "Treat a stranger like a thief" and "The nail that stands up gets hammered down," there are few advantages to making yourself exotic through foreign travel—especially if you leave the group tour and go native. American attributes especially may rankle; students returning from foreign travel were seen, in a survey of teachers, to be "smart-alecky,

weird, pushy, overly talkative and assertive, disobedient, uncoopera-
tive, and restless." Not hard to believe.

It is also not hard to believe that such a society discourages active
dissent by citizens, and puts enormous responsibility on the leadership.
Reforms that do not come from the top, often do not come.

An everyday example of a passive citizenry: a Japanese professor
friend defended whaling by quoting to me government statistics. "Do
you trust the Ministry of Fisheries [MAFF] figures?" I asked. He
seemed at once stunned, confused, and incredulous. "Oh yes," he replied. I looked equally stunned. "We never believe our government," I
said. No doubt he was thinking, "That's why you live in anarchy," and
no doubt he was partly right. Like most Japanese we met, he thought
the Western objection to killing whales was moral (whale meat as sacred
cow), rather than practical (fear of species depletion), and therefore
he saw the outside pressure as cultural arrogance, America forcing its
food taboos on Japan. That is how the Japanese ministers present the
issue; for instance, in 1987 the minister of MAFF said to Parliament:
"A large number of Japanese feel it ridiculous that it's all right to kill
cows but not whales. We feel a statement saying that it is barbaric to
eat whales is racially prejudiced."

If no distinction is made between domestic and wild animals, and
if depletion of a wild species is never mentioned, there is no debate.
Most of us agree that the situation would be quite different if Japan
were raising whales for slaughter. To this day Japanese ministers cite
minke whale populations (only recently hunted) and do not quote outside sources on other species. Many Japanese citizens believe the world
is plain wrong: there are plenty of whales, and we should mind our
own business. But few have heard the issues, and that is quite different
from debating competing sets of statistics.

A number of polls have indicated a low environmental consciousness in Japan, especially a United Nations poll placing Japanese leaders
and citizens lowest of fourteen countries in ecological awareness. It
has also been fashionable lately to attack Japan as "eco-outlaw" (*News-
week*), "Environmental Predator" (*Time*), perpetrator of "crimes against
the earth" (*Business Week*). But it is hard to generalize from polls to
public attitudes in Japan. First, many believe that Japanese, even more
than most, tend to answer polls the way they think they should be an-

swered, and second, even citizens who want to be active have a hard time doing anything in Japan. As Jonathan Holliman pointed out in *Japan Quarterly,* the problems of running a nonprofit, activist organization in Japan begin with "the virtual absence of any convenient and inexpensive way of sending money through the mail. Many public interest organizations in other countries would soon go bankrupt if they had to operate in a society virtually without personal checks, direct mail appeals, payment of membership dues by credit cards, and tax exemptions for private donations." Even door-to-door collections are regulated into infeasibility.

The government, during the Year of Environmental Diplomacy in 1989, hosted a worldwide Conference on the Global Environment. But all representatives of Japan's environmental movement were excluded (causing them to host a simultaneous counterconference), and journalist Hiroyuki Ishii said, "People all over the world were laughing about it. . . . In every country I visited, people told me that these two Japanese officials had come to see them about the conference. These officials went around saying that they had a big auditorium and a big budget and all the right facilities, but they had no idea what to discuss at the conference."

The system, in other words, does not encourage alternative thought or alternative, nongovernmental organizations. Holliman (ten years an activist in Japan), like many others, concludes: "It is more likely to be international rather than domestic pressure that will influence the government and industry to reform their environmental policies."

For the writer respectful and fond of Japan, this is a no-win situation. The Japanese are currently bearing the brunt of American sloth and arrogance, so that any critic of Japan may appear to follow a "bashing" bandwagon whose music he deplores. And yet, since so many well-informed Japanese say to one privately that the government and the corporations will respond only to outside pressure, and that internal Japanese dissent can use outside help, one should speak. This introduction, then, is an apology as well as explanation of why I find myself criticizing the actions of another country when my own, so prone to behaving badly, is uncorrected and, too often, unreproved. Moreover, as many of us predicted, official Japan is beginning to take the rational and long-range view which ecology implies: at the Earth Summit in

Rio in 1992, Japan supported far stronger environmental reforms than the United States did.

I will take the data for Japanese timber use (unless otherwise noted) from a Japanese source: Yoichi Kuroda of JATAN (Japan Tropical Forest Action Network), Tokyo, who, with François Nectoux of the World Wildlife Fund, authored a comprehensive and excellent criticism of the Japanese tropical timber trade, and who continues to bring public attention to Japan's environmental policies. "Foreign media coverage is very important," Kuroda says. He points to a Ministry of Foreign Affairs listing of negative publicity on Japan. "The government is sensitive about this."

JAPAN RAPED the Philippines. They took it all and ran. An estimated 4 percent of the Philippine forest is left. Of the many statistics we could use, consider a simple list (in the table below) of Japanese log imports from Southeast Asia from 1970 to 1989, rounded off to hundreds (the numbers represent 1,000 cubic meters). As the Philippine forest disappears, the action moves to Sabah and then to Sarawak.

Year	Philippines	Sabah	Sarawak
1970	7,500	4,000	1,900
1971	5,700	4,100	1,500
1972	5,100	5,400	1,400
1973	5,900	7,300	1,300
1974	3,900	7,000	1,000
1975	2,900	6,000	700
* 1976	1,700	8,500	1,700
1977	1,500	8,100	1,500
1978	1,600	9,200	1,500
1979	1,300	8,200	2,300
1980	1,100	6,300	2,300
1981	1,400	5,500	2,900
1982	1,300	6,400	4,000
1983	600	6,200	4,100
1984	900	5,500	4,300
1985	500	5,900	5,400
1986	260	6,000	4,800

Year	Philippines	Sabah	Sarawak
1987	27	7,000	5,500
1988	32	5,400	5,300
1989	10	4,600	5,600

*Year of the Philippine ban on logging in six provinces, and on exports.

The important variable behind the figures is simply the availability of cheap, high quality logs. Only a few major political or market events occurred during the period: the Arab oil crisis of 1974, the unenforced ban on log exports from the Philippines in 1976, the fall of Marcos in 1986, and Indonesia's decision not to export whole logs but to mill their own, keeping the jobs at home. The Japanese much prefer to import whole logs to feed their own very efficient mills, and no one really believed that Indonesia would stand up to Japan. But in the late 1970s they did, and log exports from Kalimantan (Indonesian South Borneo) to Japan, after hovering between 6 million and 8 million cubic meters during the 1970s, fell to zero by 1986. Concurrently, Indonesia made significant inroads in the sale of their own finished plywood to the Japanese market.

The Indonesian policy requires, perhaps, a strong central government, capital investment, and a willingness to defer immediate profits on timber in favor of long-range investment. It also requires some willingness to distribute wealth: local people gain jobs, adding value to the raw materials before the sale. Alas, Sarawak and the United States have hardly begun to follow the Indonesian example.

The Philippines provides the clearest picture of how the Japanese tropical timber trade operates. The fall of Marcos brought the dirty linen into view. As the new Philippine minister for natural resources said in the trade journal *Asian Timber* in 1986, the Philippine forests had been subjected to "severe exploitation and wanton plunder."

The exports to Japan began with Philippine mahogany in 1951, which was then milled and reexported as plywood to the United States for valuable hard currency. By the 1960s the internal Japanese market had boomed and the Philippine exports peaked at 8.3 million cubic meters in 1969. In the 1970s the high yield, high quality lowland dipterocarp forest was depleted and exports began to sag. Mindanao was deforested, and other provinces came on line. The Philippine logging bans of 1976, which cut the trade in half, were never fully implemented; too many fingers were in the pie. The bans were implemented in 1986, but by then little was left to save. Satellite photographs have shown much less forest left than either the government or Friends of the Earth had predicted, and it is clear that tax evasion (uncounted logs), illegal logging (uncounted logs), and illegal practices (unforeseen destruction in "selectively" logged stands) made the estimates naive. The pattern was one of "timber mining"—boom-and-bust extraction—not of forestry or tree management.

The Philippine logging for the Japanese market was carried out by local companies, and by joint ventures with Japan. The local companies, mostly Chinese, would receive a logging contract from a well-connected Filipino who had a logging concession from the government. Japanese trading houses would lend the money for equipment, repayable in logs. In joint ventures, the Japanese could directly own (in partnership with Filipinos) up to 40 percent of a logging concern. In both cases the equipment was Japanese.

Sustained yield logging, or even legal logging, was rarely part of the plan. In joint ventures, the Filipino side sometimes could not advance the 60 percent needed, so backdoor funds often accounting for as much as 30 percent of the total were supplied. These funds could not be insured. According to a Japanese observer, "As a result the Japanese side endeavoured to recover the uninsured amount as quickly as possible, contributing to the extremely rapid exploitation of concession areas."

In addition, logging concessions were for one to ten years, hardly long enough to encourage stewardship or concerns about second growth. The indirect destruction involved in such Southeast Asian logging is enormous: about 14 percent of an area is cleared for logging roads and workspace, and 40 percent of a logged area is denuded. In one "selectively logged" hectare in Sarawak where twenty-six trees were taken out for timber, thirty-three more were broken or badly damaged. Poisoning of noncommercial species is common. Sloppy forestry is not the only problem: in the Philippines, bribery of overseeing officials, illegal cutting on concessions and off, smuggling, and tax evasion were all normal.

Japanese companies were not only buying the results of such logging, they were in many cases providing capital funds and were directly involved in illegal practices. The usual skimming method as exposed by post-Marcos Philippine inquiries is simple: a timber company over-cuts its monthly quota from a concession and pays a customs official to sign two documents—an accurate shipping bill for the Japanese import officials and an understated bill for the Philippine export quota, tax, and royalty reports. A bank or front company in Hong Kong provides the neutral ground where letters of credit are exchanged and the unreported profits are shared by the three parties—the accruing capital usually staying in Hong Kong, where it is not taxed and can be used to finance other ventures with no questions asked. When I repeated this scenario to a man in Sarawak who had managed a timber camp, he laughed and said, "So you already know how they do it here. Exactly the same."

No one could seriously claim ignorance on the part of Japan. For instance, in 1984 the logs from the Philippines officially unloaded in Japan were almost twice the official Philippine export quota. Between 1984 and 1987, comparing falsified Philippine export records and accurate Japanese import records, one finds that 38 percent of the Philippine logs to Japan were illegally smuggled. The Japanese firms and government had to have been aware of the discrepancy. One must admire the Japanese thoroughness in counting. Japanese import firms have appeared content to let a few in the resource country become rich in a boom, while the import firms make plans to move on when that resource has been depleted. The Japanese sense of obligation seems to extend no further than supplying raw materials to their home industries, as a trader himself suggested in 1988:

> . . . the depletion of tropical timber resources in Southeast Asia has become a matter of reality today, so we have to look to Brazil for a new supply of resources because our trading house has an obligation to supply raw logs to plywood manufacturers. Also, this new situation would create a new bargaining power against Sabah and Sarawak and Indonesian plywood.

There is no reason to believe that the relationship between Japan, the mainly Malay ministers, and the almost exclusively Chinese timber companies in Sarawak is any better than in the Philippines. In a famous incident of 1987, the two major political parties of Sarawak had a mud-slinging campaign, revealing the timber concessions given out when each was in power. Naturally the lists abound with relatives (how *could* the former chief minister's daughter find time to serve on the board of three timber companies at once?) and with busy chief ministers' secretaries (two of whom held a large timber concession but apparently did not bother with its daily operations). In all, licenses worth about $100 million U.S. were accounted for on one side of the aisle (cronies and family of the former chief minister), a sum which will go a long way in Kuching. The generous former chief minister admitted that he had "helped a number of people plagued with financial problems to become rich overnight." In response to charges that political allegiances had been bought, he said that he did not "want his colleagues to enter politics and suffer." Ah, how that possibility is dreaded by ruling classes around the world—but the catastrophe was averted by effective government action.

The incumbent minister making these revelations, for his part, "declined to answer a question about the basis on which timber concessions were given out," and within a few days he was counterattacked, as the former government struck back with lists of its own.

Let us say that the Japanese have little to be proud of in their relations with Sarawak. They just have logs. I asked a leading minister in Sabah, and another in Sarawak, the same question: "The Japanese offer you a market for your goods. Are they also investing in your country?" The Sabah official, himself a direct beneficiary of years of cooperation with the Japanese, replied: "They have not planted a single tree. They took them all out, but they have not planted one." In Sarawak, the minister, also heavily involved in logging and its profits, said: "The Japanese are a market only. They are investing nothing. They will be finished and gone."

These men, remember, are not the opposition; they are officials in the government and directly involved in the export of timber. One very much appreciates their candor, which would be hard to match in establishment Tokyo. Yoichi Kuroda of JATAN commented in 1989:

> The main features of Japanese investment in forestry products have changed little during the last three decades. The desire for control— with no strings attached—of a given resource base, combined with the search for maximum profit yield in the short term, have acted and continue to act as the main thrust behind any investment. In-

stances where Japanese firms have contributed to the development of forestry industries in other contexts or investment policy frameworks are extremely rare.

The Philippine history is apparently being repeated, and in the interim Sabah was cut as well. Short-term licenses and "timber mining" were common in the early 1980s; now the half-private, half-government Sabah Foundation owns most of the concessions, and has introduced some control, working on sustained yield plots and reforestation. But they are shutting the gate after the horses have run; most of the timber is gone.

The world situation of the tropical timber trade is partly alarming but partly encouraging, because it shows that some tropical timber trade, especially in Sarawak, could be stopped immediately without major effects outside of Borneo. Internal Sarawakian problems could easily be mitigated by a willing first world. Even a moratorium on cutting all of the world's tropical rainforest, tomorrow, would not affect world stock markets nearly as much as the first week of Iraq's invasion of Kuwait.

Everyone knows that the tropical rainforests are in jeopardy, but just how and why varies region by region. About half of the 1.6 billion hectares of the world's tropical forests (closed, mature forest) have disappeared in the last two hundred years, and out of the half left, about one-tenth of one percent is being managed on a sustained yield basis (1987 estimate). About one percent is eliminated each year, and another one percent is grossly disrupted. One can see that there is time for significant changes of policy.

Of the remaining 1,200 million hectares of tropical forest (using now the Smithsonian 1988 estimates, slightly higher), about 220 million are in Africa, 300 million in Asia, and 680 million in Latin America. The areas of worst deforestation to date are in Asia, particularly Thailand, the Philippines, and Indonesia, and in Africa, particularly Madagascar and West Africa.

There are three major causes of deforestation, listed here in approximate order of historical appearance: (1) *Traditional shifting cultivation by fairly stable populations over a long period ("swidden" agriculture).* Usually at least three plots are thus rotated. Almost no one believes any longer that traditional shifting cultivation is a dire threat to tropical for-

ests. (2) *Logging, including roads, waste, poisoning, erosion, and water pollution.*
(3) *Cultivation by the landless poor.* This is an entirely new type of slash-and-
burn cultivation, nontraditional, created in the past few years as the
landless poor (sometimes displaced by monoculture and export plantations) rush along the new roads opened by logging, ranching, and
modernization interests. Just as farms sprang up along the railroads in
the American West, the poor are following the bulldozers into the forests. Their slash-and-burn methods are informed neither by hundreds
of years of practice nor by the stewardship that comes with a stable territory. Especially in Latin America, they pose a serious threat to the
forests. Like the fuel gatherers depleting Nepal and India, these people
are not part of "forestry" problems or solutions, but of much more
complex and disturbing problems: population, distribution of wealth,
landownership, social organization.

Surprisingly, according to Kuroda, Southeast Asia is the only place
where logging itself is the major culprit:

> . . . logging, for instance, plays a considerable role in opening for-
> ests to shifting cultivation in West Africa, but by itself is not so
> intensive that it irremediably disrupts and damages the forest. On
> the other hand, in Southeast Asia, the extent of logging activities
> is such that damage to the soil and any remaining vegetation is fre-
> quently considerable, affecting ecological cycles for very long peri-
> ods. In the Amazon basin the importance of logging activities is
> small in comparison with that of infrastructure, direct colonization
> through access roads, and cattle ranching.

Let us consider the role of Southeast Asia (the only area where logging is the crucial factor) in the world tropical timber trade. In figure 1
(showing tropical log exports, 1986), the flow from Southeast Asia
dwarfs other regions, as it does in figure 2, for sawnwood. Plywood
and veneer, in figure 3, also come mainly from Southeast Asia. In these
flows, note that about 60 percent of Southeast Asia's whole logs are
going to Japan, about 10 percent of the sawnwood, and about 13 percent of the plywood and veneer. Japan is the world's major importer
of tropical timber. In 1989, it was the major importer from Sarawak.
In 1989, Sarawak's whole logs (total of 12.3 million cubic meters) went
mostly to Japan (5.6 million), Taiwan (3.1 million), and South Korea (2.0
million). However, a Japanese trader told me that a number of the logs

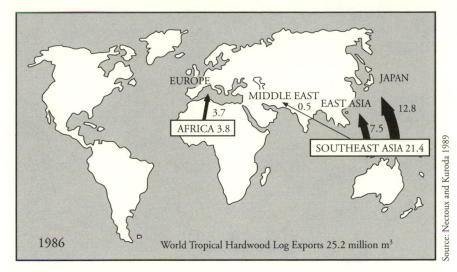

Source: Nectoux and Kuroda 1989

Figure 1.

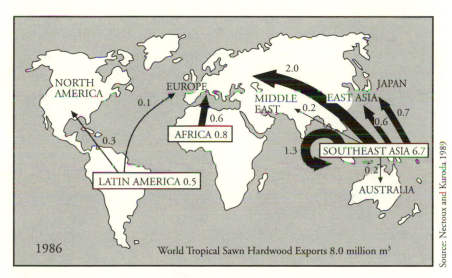

Source: Nectoux and Kuroda 1989

Figure 2.

go to Japan by way of Taiwan and Korea, so that the import statistics will not make such an easy mark for the environmentalists.

How dependent is Japan on this supply? Figure 4 shows sources: helpless Malaysia (Sarawak) sells its logs with no value added, followed closely by the hapless United States. Large timber reserves in Canada and the former USSR began to be tapped by Japan in 1991–92. According to Mitsubishi in March 1991, other major timber investment areas were New Zealand and Chile.

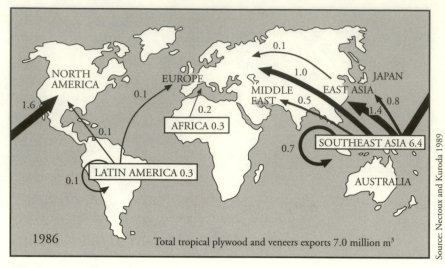

Source: Nectoux and Kuroda 1989

Figure 3.

Japanese plywood mills prefer high quality tropical hardwood, but they can and have converted mills to using softwood. Their "preference" is not a necessity. About 80 percent of the imported tropical logs go to plywood, and most of that plywood is used either in the cheap (painted) furniture business (about 30 percent) or as forms for concrete in the construction industry (about 30 percent). That is, a very significant percentage of the Sarawak forest is going to just a few end uses, open to modification, in a single country.

Some conclusions:

1. Of all the causes of forest destruction, the timber trade is the easiest to regulate. It is a way to make money, and conservation or alternative materials can modify the market demands. Southeast Asian forests, mainly in Borneo and New Guinea, are being cut for the timber trade. Compared to the circumstances in Africa and Latin America, these problems could be easily addressed.

2. The Southeast Asian timber trade, especially Sarawak's, is mainly directed to the Japanese market.

3. The Japanese have the technology, the capital, and the political sophistication to alter drastically their import patterns if they wish to do so, and to influence neighboring importers in Taiwan and South Korea.

4. The Japanese, therefore, should be held responsible for the destruction of Sarawak.

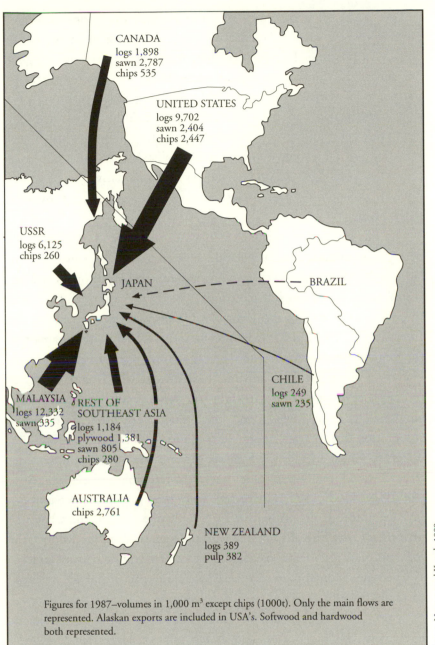

CANADA
logs 1,898
sawn 2,787
chips 535

UNITED STATES
logs 9,702
sawn 2,404
chips 2,447

USSR
logs 6,125
chips 260

JAPAN

BRAZIL

MALAYSIA
logs 12,332
sawn 335

REST OF
SOUTHEAST ASIA
logs 1,184
plywood 1,381
sawn 805
chips 280

CHILE
logs 249
sawn 235

AUSTRALIA
chips 2,761

NEW ZEALAND
logs 389
pulp 382

Figures for 1987–volumes in 1,000 m³ except chips (1000t). Only the main flows are represented. Alaskan exports are included in USA's. Softwood and hardwood both represented.

Source: Nectoux and Kuroda 1989

Figure 4. The Pacific Rim: main timber supply source for Japan

THE SARAWAK GOVERNMENT is handing out concessions and pushing in new roads to gain every legal and physical advantage as quickly as possible. In good weather, many companies are logging day and night. Since 1980, world attention to the rainforest and to the plight of natives, instead of inducing long-range planning or a slowdown, has only sped the destruction; apparently officials fear they might not get it all before they are stopped. Logging rates and actual acres logged far exceed the environmentalists' worst predictions of only a few years ago. Go upriver and ask; not one native out of a hundred will defend the current practices, even if they are directly involved in the profit.

Officials in Sarawak admit that they can see the end, and are wondering how to make money when the logs are gone. They are planning large hydroelectric projects, flooding valleys (and longhouses and ancestral tribal lands) to sell electricity, possibly to the mainland grid through a cable to Singapore. In 1991 two industry organizations, the International Tropical Timber Association (ITTO) in Yokohama and the Sarawak Timber Association, recommended a 30 percent reduction in cutting, but since the companies have now moved from the twelve-ton-per-acre lowland forest to eight-ton-per-acre hill forest, and since almost all is now concessioned, the reduction is little more than a recognition that operations should begin to shrink. Time to ship some dozers to Chile. In 1992 the Sarawak government loudly announced a 30 percent reduction in logging, but in the remaining places where the wood is good, the cut has increased. Since over 50 percent of Sarawak's revenue comes from timber, and since, off the record, the wheels of government are liberally greased with the lucrative timber licenses, the local political repercussions of timber's demise will be considerable.

The new colonial pattern is clear; although not politically enslaved to the Brookes as in 1841, or to the Japanese as in 1941, the Sarawak body politic moves beneath the hand of Japanese cash. Japanese loans, equipment, buying power, and *willingness to encourage short-term profits* offer a chance to make millions quickly, an offer which local Chinese businessmen and Malay ministers seem unable to refuse.

If you believe that the rich in rich countries sometimes work with the rich in poor countries to extract resources for a quick and short-lived profit, then you will probably find the Japan-Sarawak relationship a classic example of that first world–third world pattern. Nowhere else is our end-of-the-millennium situation so dramatically displayed: Japan —one of the highest tech, highest literacy, richest countries in the world—with its bulldozers comes face to face with "stone age savages." The savages are armed with blowpipes, a hunter-gatherer's knowledge of the forest, VCRs, and environmental defense lawyers. Their ecosystem raises worldwide concern, so that local Sarawak officials face "outside intervention" on entirely new grounds. Indeed, the idea of "sovereignty" or national independence is completely revised by the idea of global warming, with the curious result that natives fear losing control of their land to the World Wildlife Fund as well as to Mitsubishi; the new international economy impinges within twenty-four hours on the life of the longhouse; the oldest colonial issues, such as native land rights, are suddenly replayed in 1990 as if nothing had changed. No longhouses own logging concessions, although they have applied. Loss of their natural resource, the forest, drives them to economic dependence; their cultural heritage is violently disrupted as they move to town or remain behind on their ancestral land, working for wages. In a strange paradox, a renewable resource, timber, comes up against a nonrenewable resource, the rainforest, which took over one hundred million years to grow and may take a very long time to grow back.

"How much lowland rainforest is left?" I asked both the minister of the environment and the leader of SAM in September 1990. "None," both replied. Aside from the parks. That has been accomplished in twenty years.

We might remain open-minded on several fronts: trickle down theories, modernization thresholds. If, for instance, the forest is being damaged and the nomadic Penan are being forced into a new life, one could argue that their lifestyle is doomed anyway and that a country could take the enormous capital generated by the sale of timber resources and reinvest in such a way that all the people would move into a more prosperous modern age: somewhere, in a hidden silicon valley, happy Penan assemble microchips. Fujitsu-la.

Unfortunately, even the pro-logging Chinese in Kuching say that most of the money is going out of Sarawak to private accounts in Hong Kong, while taxes and funds for mitigation of timber consequences are minimal. I found nearly universal agreement on this point. Theories of modernization can provide an effective counterpoise to European romanticism, but the actual practice must be carefully monitored: Is the capital being reinvested in the country? Are the natives free to choose to log or not to log? Do they have any control over the production of timber on their native lands, or over the money generated by its sale?

And finally, to anyone who has been there, another way of looking at things: the oldest and richest and most delicate beauty on earth—the rainforest—is being destroyed for the oldest, ugliest, and simplest reason: greed.

The situation, however, is not hopeless. Most of the forest may be concessioned, but it is not all cut. Much upland forest is uncut; moreover, two cuttings damage but do not destroy the upland ecosystem. Wood *is* renewable, and logging could be done well. By whom, and how fast, and who profits—those are the issues. And, where is that wood going?

CHAPTER 7 🦎
TOKYO RISING

In the predawn light, the streets of Tokyo are empty. The trains and subways stopped just after midnight, and the last drunk salaryman has long ago stolen a bicycle and wobbled home. It is still half an hour before the subways will start again. Isolated footsteps echo in the various downtowns—Ikebukuro, Shinjuku, Shibuya, the Ginza—islands of tall, shiny office buildings and huge department stores rising here and there from a sea of neighborhoods made of houses and shops, a sea of humanity, thirteen million people, stretching for miles. Take the train in any direction for an hour—it will look the same. Within an hour and a half you are in the suburbs, but they too are a seamless web of buildings, once distinct towns, now solid shops and residences broken only by occasional tennis courts, by huge golf driving cages with green nets a hundred feet high, and by tax-dodge carrot fields where here and there some speculator waits, holding his "agricultural land" against even higher prices.

Down at the docks, at the Tsukiji fish market, the city awakens. Far before first light the boats are unloading at the piers, and wholesalers begin to shout their orders and direct trucks. Soon the dock is alive and the action spreads inward to the retail stalls: five football fields of fish sellers jammed under one roof, bins and tables overflowing with the still flopping and twitching catch, ruddy and energetic and good natured people as far as the eye can see crammed behind and between the tables, moving, shouting, telling stories, practicing golf swings and laughing at 4:30 every morning, selling the day's fish to Tokyo, Yokohama, and the environs, to perhaps fifteen or twenty million fish-loving people. Hardly a bite will be kept overnight. In Japan, everything must be quality; everything must be new.

Soon the Yamanote line rumbles through on its overhead tracks, the ten subway lines deliver their first passengers, and the traffic begins to pack the streets. By 6:30 a.m. the rush is on, and cars soon come to a halt. You will hear no horns, however; their use was outlawed except in emergencies. The day after the law was passed the city was quiet. An identical law in New Delhi, Bangkok, Paris, London, or New York would have little or no effect, and vast numbers of police could never enforce it. In Japan, that law was obeyed. The streets you walk near the market at dawn are clean—five cigarette butts and two candy wrappers on the sidewalk in one block—and safe. This is the only modern, industrial city in the world that works.

Rays of sun catch the tops of buildings in the Ginza—the black Sony tower, the Matsuya and Mitsukoshi department stores, and the banks—then the empty blocks where the ministries sit in the dull, massive architecture of power that neither capitalism, communism, nor Japanism has improved: the Ministry of Foreign Affairs, the Ministry of International Trade and Industry (the infamous MITI that has been "accused" of intelligently directing Japan's investment strategies, an accusation that only Americans could make without chagrin).

And above, across the moat from the Imperial Palace, the Mitsubishi blocks, once the Imperial Army's marching grounds: the Mitsubishi Shoji building, the Mitsubishi Electric building, the Mitsubishi Heavy Industries building, the Mitsubishi building, the Mitsubishi bank. Within a stone's throw are the Mitsubishi Trust building, the Industrial Bank of Japan, the Bank of Tokyo, the Hong Kong and Shanghai Bank, and the Imperial Theatre.

Tokyo has been leveled twice in this century, once by the earthquake and fire of 1923, and once by American bombs and fires in 1944–45. In 1946, very little was standing in strategic areas of Tokyo. Many other cities, besides Hiroshima and Nagasaki, were heavily bombed. Only the great cultural centers of Kyoto-Nara and Kamakura were off-limits to American flyers. After the war, the construction industry boomed.

By the mid-1950s it was as if an anthill had been kicked apart; cities seemed to rise again in hours. From 1955 to 1973 the amount of timber consumed by Japan doubled. The doubling depended entirely on timber imports, as contributions from domestic forests actually declined. In 1955, the Japanese imported almost no timber at all; by 1973 they imported 70 percent of their needs. Since the oil crisis of 1974 the market has been erratic, with several significant housing and timber slumps in the early 1970s and early 1980s (some plywood mills folded in the eighties). Japanese domestic timber has become increasingly expensive and specialized, for high quality uses, while construction styles and therefore wood demands have changed, and tropical timber is becoming more rare and expensive. Change is in the air. Industry leaders, as well as environmentalists, do not really expect business as usual to continue much longer.

Like Americans, the Japanese love wood in the home. As we have a soft spot in our hearts for log cabins, they revere traditional wood architecture. The effect is stunning: a bare room made of blond wood panels, white paper screens and straw-colored tatami floor mats, set off, perhaps, by a single green vase with two bare branches and a spray of pink flowers. Perhaps as many in Japan now own such a room as Americans own log cabins. But wood, like rice, is part of their culture.

Anyone who knows the rich colors of tropical hardwood can well imagine its appeal to the Japanese; such aesthetic and irreplaceable use, however, is not destroying the forests of Sarawak. For such special uses, Sarawak's wealth of unique wood could be harvested forever.

Plywood consumes 60 to 80 percent of the tropical hardwood entering Japan, and most of that plywood, used for both cheap furniture and concrete forms, is soon thrown away.

On first hearing this fact, many are stunned. The rainforests are being cut to make plywood? Precious tropical timber with its exquisite grains, mahogany, teak . . . It is hard to believe that a unique and beautiful resource would be used for a faceless task, just because it is a few cents cheaper than pine; it is hard to believe that no one is looking ahead; it is hard to believe that Japan, justly renowned for restraint, economy, purity of style and action, could be so wasteful.

One problem is indeed Japanese aesthetics: tropical plywood has no knotholes, and once accustomed to its smooth perfection, the Japanese are loath to use inferior substitutes. The custom began, of course, as Japan sought nearby forests after the war, and they originally used Philippine hardwoods to crack the American plywood market with a tropical import. Now many observers in Japan lament such use of tropical wood, or the use of any wood in *kon-pane* forms for cement; but the construction industry, largest donor to the Liberal Democratic Party (LDP), is moving too fast to change habits unless forced. Many substitutes for wood in *kon-pane* are possible, and are now used by some large firms (such as Ohbayashi) for standard contracts; various kinds of concrete work requiring special forms could easily use pine, fir, and other monoculture and plantation trees. As labor becomes more expensive, companies are looking more at precast concrete, metal, and masonry forms. In short, a world moratorium on tropical cutting would simply force a shift in construction techniques, a shift which all know is coming anyway in five to twenty years. It would force that shift before the rainforests are gone, instead of after.

Might this happen? There are many sides of most questions, and if Japan can be said to have little dissent and a passive citizenry, it can also be said to have some of the most effective leadership in the world. Official Japan is capable of change, with a swift efficiency that puts other countries to shame. After all, over 80 percent of the upper ministers are graduates of Tokyo University law school, and when they get on the phone to each other and to their undergraduate Tokyo University classmates who now lead the larger corporations, things happen.

Some of us believe that the Japanese could easily, in a few years, lead the environmental movement. They are smart enough to see that rational self-interest depends on a worldwide caring for resources, and they are capable of the long-range planning and technology that helps us adapt to a new age. If Japan comes to believe in what much of America and Europe are saying, they may become the best practitioners of environmental conservation.

Many have pointed out that *within* their country the Japanese have already demonstrated these truths. In the 1970s, they led the world in turning around pollution, especially exhaust emissions, in Tokyo. In 1990, they began to cooperate on endangered species, and went from being the world's largest importer of ivory to virtually banning its use. "I think the top leadership in MITI had had enough," said Tom Milliken, an environmentalist in Japan. "Japan has taken a real beating on this in the local and foreign media, and they decided it just wasn't worth it, having Japan's name blacklisted constantly, to protect one small industry. They decided to cut their losses."

The Japanese have signed the worldwide ban on CFC (chlorofluorocarbon) production, crucial to microchips, and have committed themselves to developing alternatives. (Singapore claims, however, that this will simply keep Japan in control of Asian microchip production.) Their energy efficiency leads the world (Japan uses about half the oil per production unit—car, TV, whatever—as the United States), and

Tokyo is planning an elaborate central heat supply system to capture waste energy from factories, subways, and other facilities. The Japanese have grudgingly modified their whaling position, and among their self-interest forest aid plans (mostly to start plantations) is a fine project to help Southeast Asian residents identify marketable forest resources such as moss, ferns, rattan, medicinal plants, perfumes, and dyes, "contrasting with the policy the government has hitherto pursued of promoting felling and replanting activities for foreign trade earnings," as JATAN observes, Many think the Sarawak forest could be renewably "mined" for many resources, if we would leave it standing. The legislature's environment committee attracts five to ten times as many participants as it did a few years ago. In 1991, the government pledged large sums to solving environmental problems. In May 1991, I asked Takako Doi, then the leader of the opposition, if the LDP's new "green" attitude is public relations or a genuine shift in policy. She, no lover of the LDP, replied that it is a genuine shift, that the leadership is changing its mind on the environment.

It would be well to remember, however, that Japan sometimes moves only when forced. The classic and shocking example is that of the Minamata poisonings, which gave birth to the environmental movement in Japan. In the words of Karel van Wolferen (*The Enigma of Japanese Power*, pp. 55–56):

Among the post-war pressure groups that have stayed outside the System, the most successful—measured by the extent to which the System has had to accommodate them—are those that have fought on behalf of the victims of industrial poisoning. Even so, it took some ten years before pollution problems received serious consideration in Japan. Only when cats that had eaten mercury-laden fish began to have spasms and jump into the sea did the problems of Minamata (the municipality that was later to become a symbol of environmental pollution the world over) first attract some limited attention from the press. Several years more passed before the plight of the human victims became a national issue. They had been rapidly increasing in number, dying sometimes of sheer exhaustion as a result of the spasms caused by their injured nervous systems.

The corporation responsible for dumping mercury waste into the sea maintained that there was no connection, and hired gangsters to manhandle petitioning victims and their families. The victims themselves, mainly from poor villages, were ostracised. The bringers of bad tidings—doctors who were studying the Minamata cases and similar mercury-poisoning cases in Niigata—were at first discredited. The Kumamoto University research team saw their findings suppressed and research money cut off. A campaign was waged against a certain Dr. Hagino Noboru who was treating patients in Toyama prefecture for the so-called *itai-itai* ("ouch, ouch") disease, a mysterious affliction in which bones became so brittle, owing to cadmium that had been allowed to seep into rice paddies, that they fractured at many places. The suppression of evidence and the hiring of doctors to claim lack of scientific grounds for the complaints were repeated in other parts of Japan until riots, the storming of the factory responsible for the mercury waste in Minamata, a national press campaign and foreign publicity finally made such approaches untenable.

The first lawsuit on behalf of pollution victims, filed in 1967 by leftist lawyers, was quickly followed by three more. In the early 1970s, prodded by this legal action and popular discontent with the badly polluted air of the capital region, officialdom, together with some politicians and businessmen, concluded that some measures had become inevitable. Strict industrial regulations were introduced; overnight, it became fashionable to decry industrial expansion achieved at the expense of the living environment. Courts of Justice, too, though they took ten or more years to conclude trials, were beginning to award damages to victims. . . . The System had responded to pressure caused by an outcry greatly amplified by the press. Its response was evidence that it can be moved to right wrongs, provided the wrongs are sufficiently eye-catching and can awaken widespread public indignation.

The problem is that as Japan has solved its own environmental crises, giving citizens a sense of well-being, it has exported its pollution to the third world, where it goes unnoticed by those at home— those who might exert some pressure on the corporations. In one famous case, Mitsubishi Chemical had moved a fluoride operation to Malaysia, where radioactive leakage was said to be causing leukemia, and certainly was causing demonstrations in front of the Japanese embassy.

Japan is soaking up East Asia's resources. It uses less wood per capita than the United States and Canada, but those countries grow their own trees on vast land masses. Japan uses seven times more wood per capita than Europe. Those who have lived in Japan see the waste of a new affluence increasing rapidly, with no end in sight. Many are puzzled; they think of the Japanese as frugal, and as lovers of nature. How, then, can they be "environmental terrorists"?

The waste in Japan is an interesting issue. Their use of paper and packaging for purchases, groceries, and gifts is legendary (three wraps around an item—paper, bag, bag—are not unusual). Packaging customs, however, might be called traditional, part of being neat and kind. More surprising is their wholesale adoption of American habits: buy it, use it, throw it away. Buy another. The cheap plywood furniture made of tropical timber is regularly discarded after a few years of use; Japanese garbage bins can be hard to distinguish from sidewalk sales, and many pilfer quality items before the truck arrives: tables, chairs, working TVs. At ski areas, few have skis older than three seasons (we know skis; our daughter is a ski instructor), and few of those we talked to at five resorts ski more than three weekends a year. Yet Japanese friends, after three hours on the phone, could find no second-hand ski market in Tokyo (thirteen million people, over a hundred ski areas within weekend distance). This means that a lot of $400 skis used ten times are thrown away. The used car market, too, is minuscule. Everything must be new.

The Tokyo Motor Show in November 1990 attracted over 2 million visitors, 300,000 in a day. Three women in skintight leotards exercised "motorcycle muscles" at the Kawasaki display; bathing suits with legs and high heels lounged on hoods; a Honda executive in suit and tie wiped fingerprints. "Everyone must contribute to the show," he said. The displays of every major manufacturer from Japan, England, Germany, Italy, and America covered acres; all displays were built on platforms made of tropical timber plywood; all of it was thrown away and burned. One manufacturer threw out 200 sheets of plywood; another 980. The twelve-day show produced 383 tons of waste.

Time magazine of July 10, 1989, reported the conventional wisdom on Japan: "The country's extensive program of garbage recycling is a model for all industrial nations." We thought so too, in 1990 in a Tokyo suburb (Asaka-shi), as we carefully sorted our waste: plastic bottles, glass, metal, and aluminum. All were taken out one day a week and placed in the appropriate containers; usually a housewife was standing near correcting any mistakes. A can replaced here, a bottle cap there.

Then one day Juliette watched the men come, pick up the garbage bins, one by one, and empty them into the truck. All together. Not long after, she visited the Asaka incinerator with Japanese friends, housewives interested in environmental reform. In the control room, men in beige uniforms with white gloves sat in front of computer consoles and video screens, overseeing the latest German equipment. Through her friends, Juliette asked the manager how they keep the recyclables separate, since they are dumped together into the truck. "They try to throw the cans to the back," he replied, somewhat lamely. Through the control room window, she could see cans, bottles, trash, everything going into one incinerator. "But you are burning everything together. Why do you have us separate it?" Her friends and interpreters, the most informed, activist women we knew in Japan, were a bit discomfited by the direct challenge to authority, but they translated. The reply: "We are training the people for the time when we will recycle." Said one of the ladies, "I've been separating my garbage for twenty years!" None had ever had the slightest idea that it made no difference. "Actually," volunteered the manager, "the aluminum cans are a problem. They burn too hot for the furnace." This program of garbage recycling may indeed be a model for industrial nations: propaganda.

It seems especially difficult to reconcile environmental destruction with the Japanese love of nature. Many have said that the Japanese have no word for wilderness, but it is not true. There are words for wild nature, for improved or farmed nature, for parks, etcetera. The concept of wilderness as valuable, however, is European, and can be rare in Asia. A friend of ours took two Chinese students to a nearby wilderness in Montana, and they came upon a band of mountain sheep. Our friend whispered excitedly that the animals were wild, not owned by anyone, not used. The Chinese students understood, but they were not excited, not even impressed. So what? They have a word for wild, but they do not have our relation to it. It is not sacred.

In Japan, someone speaking of "nature" and "love of nature" usually refers to improved nature: the plum and cherry blossoms on trees planted just so, between the house and the mountain, or across the lake in a park. The assumed split between man and nature is a European obsession; wild nature as a corrective to evil civilization is a European idea. In Japan, a civilized nature is by far the more important ideal. The great parks in Japan are moving testimonials to a harmony of man and nature: trees, shrubs, and flowers are placed in a sculpted landscape to create controlled successions of colors and blooms, from certain points of view, in various seasons, in various centuries. Yes, the gardener must anticipate when the five-hundred-year-old tree will fail, and where another should be fully mature in two hundred years. Sometimes this gardener is a direct descendant of the original planter. For centuries, the family of gardeners, and the garden, have grown together.

An awareness of a "park" or "garden" nature is deep in the Japanese psyche; traditional poems establish in the first line a place and season; traditional theater does the same; even kimono prints are adjusted to season. "To be in Tokyo in April is to view the cherry blossoms . . ." began a flyer of 1990. Flower arrangement may still be the most widespread hobby for young girls to study.

But from the "wild" American—especially Montanan—point of view, there is a dark side to the Japanese love of nature. The bonsai tree, constricted by metal coils, twisted, pruned, shaped to the desire of the gardener, stands like a living image of suppression. One thinks of the obedience and conformity demanded in Japanese schools. Human domination of the environment is always potentially ominous to us; we do it, but we feel ambivalent; much less so in Japan.

Down the road, this may become one of the more important issues on earth. The great divide is approaching: we either preserve the species variation left, the "wild" or evolved nature we have inherited, in the belief that it can help us or teach us something or even has its own rights, or we count on our own ingenuity to replace or improve on nature. We are doing both now, preserving here, splicing genes there, but soon world population will force a confrontation: do we need nature, or just better science?

The Japanese, with their enormous economy and advanced technology, will arrive at that crossroads with great influence, and there is every indication that Japanese citizens as well as leaders are quite willing to accept both managed nature and synthetic substitutes for nature.

Among even our closest Japanese friends, we failed to find a sense of irony when they returned from Hawaii to say no, they had not been to the beach; the hotel pool was fine. It was cleaner, after all, and safer. While we were in Tokyo, two new giant pool complexes opened; on the front page of the *Daily Yomiuri* was a picture of three surfers on a wave; they looked pretty good. We did a double take, however, reading the caption:

> Surfers show off their skills at the Tobu Super Pool . . . which opened early this month. . . . Japan's million or so surfers, said to be discouraged by the lack of big waves along Japanese coasts, sometimes go all the way to Hawaii to upgrade their skills. At the pool, however, they can ride 2-meter-high waves between 6 p.m. and 9 p.m. during "Surfer Time." Japan's heavy industry firms, looking for profitable new businesses, are keen to enter this burgeoning new market. NKK Corp., a major steelmaker, also plans to construct a surfer pool in Tsurumi-ku, Yokohama, by the summer of 1992.

Such pools are not simply imposed on a passive populace by big businesses which have polluted the coast; traditional Japanese culture is much more disposed than ours to enjoy such a man-made "nature." These issues are immediately important in addressing the problems of Sarawak: if Americans lean on the Japanese to save the rainforest, we should not assume that a wilderness designation springs first to their minds, or to any Asian mind, as a proper alternative to rapacious development. They think first of managed nature, a garden. Respecting their culture, we have some responsibility to think of alternative uses of forest products, and ways to give natives control of a progressive economy. Beyond our shores, leaving it alone is not so powerful an argument.

Lastly, Japan has the same problems with fads as the United States. How serious is the new environmentalism? Subway posters advise, in English: "Look at Nature." How long will the movement last? A Tokyo newspaper caption, beneath a picture of two pretty girls in modish haircuts: "Models sport new hairstyles designed for this fall and winter by Mitsumasa Taniguchi, at the New Hair Mode Show. . . . The variety of short cuts featured at the show reflect interest in current global issues such as ecology and environmental protection, Taniguchi said."

Echoes of H. L. Mencken: "No one ever went broke underestimating the taste of the American public."

One can conclude that consumption in Japan is a real problem, as it is in the United States, and likely to become worse. The third world's belief that reform begins on *our* doorstep, not theirs, is well founded. Moreover, we cannot expect Asians in either Japan or Malaysia to come to the rainforest with the same assumptions as Americans. We have common ground, however, in the obvious and immanent limits to our present practices. In addition, the entire world shares an interest in the claims of small and local business versus conglomerate business, which aims only at export. Increasingly, small and large businesses do not compete for the same market. They imply entirely different economies (varieties of small harvests, for instance, nuts, fruits, condiments for rice, versus monoculture plantations of oil palms of no use locally); they imply, also, entirely different control of those economies. In these concerns, Japanese citizens, American citizens, and Sarawak citizens have much in common. It may be that years of cold war, teaching us to think in terms of pro-capitalism or anti-capitalism, diverted attention from the more serious problem: small business and local control, or big business and central control? No theory frees us from the tendency of power to coalesce.

VERY SOON THE rising sun hits the unfinished tower of the new Tokyo Metropolitan Government office building in Shinjuku. It is the largest building in Japan, in the new downtown being built to rival the Ginza as a center of world power.

At 7 a.m., workers gather outside the locked construction gates. They are a strange sight in this city, for they are different. In Japan, any face that is too dark or too light, that is framed by curly or blond or reddish hair, leaps out of the crowd; something in each one of us exclaims: My God, he's not *Japanese!* After a while, even the mirror is a shock. These men are Korean, Thai, Vietnamese, Chinese, Indian, Iranian, Turkish, Malaysian. Their numbers are increasing. The Japanese no longer want certain jobs; they are beginning to wink at the few legal and many illegal foreign workers entering. These men will work for almost nothing, live in tiny (six by twelve feet) apartments in Ikebukuro, and send back sums immense in their homelands. Each morning the Japanese job bosses for a hundred different jobs at a hundred different construction sites will hire their workers for the day out of these crowds of hopefuls.

One group of eight men in T-shirts and work pants is assigned a duty; at thirty seconds past eight o'clock they pull on their gloves and come forward to meet the first delivery of the day: a truckload of plywood from a mill near Yokohama. The Shinjuku building will use 470,000 square meters of throwaway plywood forms, which means about 200,000 sheets of four- by eight-foot plywood, which means about five thousand trees, or two to three thousand acres of hill forest in Sarawak, belonging to one longhouse or another. These men, from around the world, have no idea, as they stand in lines at the rear of the truck, that they are passing back sheet after sheet of Richard's tree, from the tribal lands of Long San, in the middle Baram, in Borneo.

After two or three uses, within a year, the plywood panels will be ruined by water, by the acid in the concrete, and by being ripped out of the positions in which they were nailed. Then the remains of our meranti tree, that massive red trunk once encased in a tangle of vines, moss, and flowers—perhaps the bird's nest fern with its orchids still lies next to the stump—will be thrown away in the huge landfill in Tokyo Bay, or trucked to Chiba prefecture and burned.

CHAPTER 8 🦂
JAMES WONG, JAPAN, AND
THE 1987 BLOCKADES

In the 1930s, in a sleepy village on the Sarawak coast, a rubber planter, William Wong, sat typing one of his letters: "No joke, child, three katties of fresh fish only once go round, and each plate just a wee bit. No joke this, and we must think correctly and act correctly. . . . I shall pull James out [of school] before long and put him to study business and planting and in workshop to qualify him to take my place. . . . No joke to take my place—one must know quite a lot and have some grit and courage."

And thus he wrote to his son James, in school in Kuching: ". . . then you must come home and be Ko Ko [Older Brother] of the Home and plant something for the future. You will have to learn much arithmetic, little surveying, chemistry and bookkeeping, and carpentering, and a little engineering &c. No joke, James, to be a man. Not easy at all."

William Wong was one of many Chinese immigrants who came to Borneo at the turn of the century, worked hard, but never got ahead the way he had hoped—just like most of the settlers on America's agricultural frontier at the same turn of the same century. He remained poor all of his life. He and his wife raised eleven children, without ever thinking that any would become rich, or that he—or his son James— would come to know the Japanese.

His letters have been lovingly edited and introduced by his son James, the pioneer of hillside logging in Borneo, owner of one of the largest timber concessions, and the Sarawak Minister of Environment and Recreation. James, a vivid writer, introduces his father's world of North Borneo in 1935:

Cast your mind back half a century and—unless you happen to live in Borneo—across the seas to a tiny town named Limbang, in northern Sarawak, standing beside a river of the same name which flows into Brunei Bay. A mile and a half from the town, in a wooden walled, thatch roofed house, a small, balding man wearing spectacles, clad only in a sarong and singlet, adjusts a kerosene lamp in the tiny room he calls his office. The sun will not rise for another two hours and the countryside is dark and quiet. The man rolls a sheet of paper into a portable Underwood typewriter, takes a sip of coffee, lights a cigarette and begins typing with surprising skill and speed. Children sleeping in adjacent rooms do not stir, so accustomed are they to this pre-dawn noise. The typist completes a couple of business letters, sips more coffee, starts another cigarette and begins a new sheet. Here is what he types:

I shall begin from now to write to my Children and those who come after them things that I now consider interesting judged in the light of the present day; things that will be of benefit to them; things that took me a long time to know; things that I am practising with great benefits; things that may be very instructive to them. . . . I write in as simple a style as I can, for it is the only style I can write fairly well. . . .

So help me God.

T. E. Wm Wong, 21/6/35, 6 am.

Dogs bark, cocks crow, jungle birds set up a dawn chorus, smoke from cooking fires rises in the still air. Very soon the man who signs himself T. E. Wm Wong is embarked on a hard day's work. . . .

He was born in China, adopted by his father's brother, brought to Borneo while still a toddler [in response to an appeal by the British North Borneo Company for immigrants], became in effect an orphan, and was rescued from life probably as a coolie by the Church of England's Society for the Propagation of the Gospel, which was very active among Hakka Chinese immigrants in Borneo. ["Life in China is misery I am told, as in those days a teacher would get only $30/$40 a year and labourers $10/$20 a year plus board. That is what made people run to the South Seas."] He went to school in Sandakan in British North Borneo (now the Malaysian state of Sabah) and in Hongkong, became a teacher in mission schools, then a clerical worker on rubber estates, and finally a small rubber planter in his own right, in Limbang, in Sarawak.

There is no doubt his hard and largely loveless childhood left a lasting impression on him. Though he rarely spoke of it directly, he was fond of a poem in Hakka about "a little boy with no father, no mother, no brothers, no sisters, no cousins, no kin, without rice-fields or land, all alone."

In the 1930's Sarawak and Brunei were still without airmail services and letters and newspapers took a week or more to arrive from Singapore. Radios began to make their appearance during that decade, however, and we were able to listen to world news from the BBC and from Manila, in the Philippines.

The town [Limbang] had few amenities . . . but it did have a club. . . . Papa would visit the club several times a week to chat with friends in the evening and to read newspapers there, in particular the Straits Budget, a weekly compilation of news from the daily Straits Times in Singapore.

I remember one evening when something he had been reading there caught his attention. He called me over, slapped me on the back and said, "James, James, great news!"

"What is it, Papa?" I asked.

"Oh, my son, Rutherford has finally split the atom."

"How did he do it?" I asked.

"With a cyclotron," Papa said happily. . . .

Our daily routine during that period began with morning exercises followed by morning prayers with Papa at the organ, and with a music lesson.

Then it was breakfast and work. Each of us was given a "contract," a quota of work to do each day, such as clearing lallang — tough, long grass — from around pineapple or banana plants, or preparing a plot of ground for planting vegetables, or some such task. . . .

Papa would have a nap in the afternoon — deservedly, since he worked harder than anyone else and since he normally got up about 3 a.m. or 4 a.m. to work in his office — and then we did gardening from 4:30 till dark. In our spare time, we drifted into the library and read the books from Papa's considerable collection (which during the Japanese occupation, according to a list he typed out then, amounted to about 300 volumes). Books I remember studying included the Old Testament, Treasure Island and Winston Churchill's The World Crisis.

James Wong grew up outside Limbang, a "one street town with five rows of shophouses" just above the Baram in north Borneo, in the wake of his 125-pound dynamo of a father who, unlike other Chinese, preached "the virtues of raw vegetables and fruit" and attended all family births and many outside, sterilizing the equipment and cutting the cord himself—his firstborn having died of an umbilical infection. He had a collection of muscle-building exercise grips and pipes, and believed in the children stripping in the house: "'Don't be ashamed,' Papa would say. 'Nakedness is beauty.'"

William Wong had the first outboard motor in Brunei Bay, bought a radio and BSA motorcycle—"the first thing he did was to take it to pieces to learn how it worked"—and was not above practicing his own brand of Chinese-native folk medicine. He wrote elaborate reports on his agricultural experiments, challenging several assumptions about appropriate produce in Sarawak, and, ever on the lookout for the independent rubber planters, he fired off letters to the Government Secretary: "What is the State Bank doing? I am told before its inception when I was young that its chief function was to encourage and support industries; and I suppose it has given lots of loans to big concerns to absorb smaller ones."

He had periodic health checkups—he, of course, being the consulting physician. After taking all his measurements and considering things rather thoroughly, he would write himself a report. On June 1, 1921, "before going on leave in order to compare afterwards the effect of Holiday, Change, and Pelmanism [mind-training]," he wrote:

> Health very bad, weak, sexual powers at its decline, Strength gone, feel tired and knee numbness. . . . Morality indifferent. Fond of sleeping. . . . Careless in work and indifferent honesty. Lewdness, slight initiative and commandful. . . . Quick ejaculation. Drinks a bit. . . . Very quick in figures. . . .

We are not surprised to learn that a holiday and change helped. By 1923 he was much improved:

> Health very good and strong. Sexual powers normal and strong. Appetite very good . . . Openmindedness . . . Straightforward. Fear of God in all things though still wicked in sexuality. Greatest sport in fishing. Drinks somewhat. . . . Hate rogues. Very liberal and gen-

erous and considerate which together with trusting too much on friends and relatives ruins me financially. Found out that I could not have a better wife than present one. 3 children. . . . Not so selfish and mindwandering. Memory improved. Ejaculation normal not quick. . . .

T. E. William Wong

10.2.23

Eleven years later, the middle-aged man found a few changes:

Report made of oneself on 25th Nov 1934 on Eve of being Free from the slavery of debts, after having builded and founded an Estate, called Limbang Estate, of 516 acres of rubber with 30 acres of other detachable properties. Age 45 English, Children 8 and 1 coming. . . . Health fairly good and fairly strong. Sexual powers on slight decline and fairly strong. . . . Not quite openmindedness in fact lost the meaning of this word. . . . Not quite straightforward tempered by experience and mellowness. Fear of God not so much but belief in Jesus Christ more. Mellower and not quite so easily provocated except brood and feel hot, at times very hot indeed at being taken advantage of unduly by present partner whose money did nothing during severest world slump except growing fat at interests. . . . Fond of drinking beer and whiskey. . . . On Religion, found that in no matter what occupation and positions one is placed, do that work with all your heart, your soul, your mind, one is actually doing God's work.

To one of his daughters' suitors, this most candid of patriarchs gave unsolicited encouragement on "the Art of Wooing. When I was your age before, Mr Tan, the shyest bird perched on a tree would come down to my coo-cooing!"

Then there were those homely occasions when we stood around the organ at home while Papa played and we had a great sing-song. Papa was a fine organist—he played for years in churches in Sandakan and Kuching—and had a rich voice as well. . . . favourites included "Silver Threads among the Gold," "Sweet Marie," "Gipsy's Warning" and "Beautiful Dreamer," as well as a Malay song, "Fly, Little Fly, Settle on my Toe." . . . Papa liked to describe these family get togethers as "thy wife like a fruitful wine & thy children like olive branches round about thy table."

ALTHOUGH THE WORLD was not counting its population in 1890, when William Wong Tsap En came to Borneo as an infant, it was certainly feeling the effects of the doubling of population from the year 1 to 1500, and again from 1500 to 1800, and again from 1800 to 1950. By the 1890s, Europeans were rushing to the American West just as Chinese also were rushing there and to Southeast Asia and to anywhere else they could book cheap passage. America's 1886 Statue of Liberty, recently duplicated in China's Tiananmen Square, seems to have been equally appropriate to either the Atlantic or Pacific predicament: "Give me your tired, your poor, your huddled masses yearning to breathe free." One out of every five people on earth is Chinese.

Four main groups of Chinese came to Borneo. Wong's group, the Hakka, were the most numerous. As in the American West, the men came first as miners, then more and more families came as cash croppers around Kuching. The second largest group, the people from Foochow (Fuzhou), settled mainly along the lower Rajang, from Sibu to the coast. Hokkien came mostly to Kuching, and Teochiu also preferred the towns. These groups, and others, spoke mutually unintelligible dialects of Chinese, and typically formed their own colonies, neighborhoods, temples, and societies. They did, and still do, educate their children as carefully as did William Wong, in community Chinese schools. Today every Chinese child will speak Chinese, Malay, and English—very well. Success as an immigrant is an ideal; assimilation is not. According to historian Daniel Chew, they also tend to gather in their own occupations: the Hainanese start coffee shops and restaurants; the Hupeinese are "tooth-artists"; Chinese from Kiangsi (Jiangxi) are expert furniture makers.

When James Brooke arrived in 1839, the Chinese had a thriving mining district in the mountains bordering Kalimantan. They had never come under the direct supervision of Malay rule; the Malay princes were content to exact tribute from the Chinese at the downriver ports. Brooke, with typically English aspirations to dominate a territory, took the Chinese on as soon as possible, by force. Over a number of years, laws, and raids, he brought the mining under Kuching control. In 1857 the Chinese rebelled and burned the palace in Kuching. Brooke soon retook the town. The rebellion was a direct result of Brooke's interference in what the Chinese considered their own autonomous mining region. Similarly, as Hakka turned to cash cropping pepper and other produce, the Brooke government vigorously asserted the right to regulate and tax. The Brooke-Chinese antipathies were of long duration, and tied directly to British colonial behavior in Asia in general; tied, also, to the threat of serious economic competition from the industrious and educated Chinese, a threat the natives did not pose. To this day, the Chinese-Brooke competition colors the Sarawak Chinese response to English-speaking environmentalists, who are easily perceived as anti-development, anti-Chinese, pro-native.

The Chinese competed also with the natives—for land. They sought permanent, individual ownership—anathema to native custom—and their cash-crop plots were often in native districts. According to Chew, in his excellent book *Chinese Pioneers on the Sarawak Frontier*, the Hakka gardeners were excluded from mining interests as the British gained control of the mines, while natives controlled most of the agricultural land. Naturally, the aggressive Chinese immigrants pushed against the weaker adversary. By 1887 the *Sarawak Gazette* was reporting: "Land disputes and collision between Chinese gardeners and Dayaks are somewhat constant, and from the number of pepper gardens opening all around Lundu, such disputes are bound to increase." This is quite similar to land conflicts, in the same years, between white settlers in the American West and Native Americans.

So when contemporary Chinese in Kuching do not want to hear more about the Brookes, and seem a bit testy about native land rights, they are acting out old traditions. Chew (himself Chinese) points out that those from Foochow, especially, moved from rubber planting into trade, banking, real estate, and timber. By the early twentieth century, he says,

> the Foochows' requirement for land for permanent cultivation, their feelings of insecurity, and their desire to constantly expand and protect their economic niche without due regard for native land rights brought them into conflict with the Ibans. In recent years, for example in 1987, conflict has again occurred over land-related issues— this time, the demands of the timber-logging industry versus indigenous land rights. The present conflict is more complicated than previous conflicts in that the present timber concessions are held not only by Chinese but also by élite groups from other communities.

Still, this does not negate the point that it is instructive to understand the present timber-logging disputes in Sarawak by examining the earlier land disputes in the Lower Rejang at the turn of the century. It may be revealing to note that some present-day attitudes towards land use may not differ from attitudes held earlier in the century. For example, it is still believed that shifting cultivation as practiced by the natives is "wasteful" as compared to the "usage of the land on a more permanent basis" by the Chinese.

From the Chinese point of view, native tribal customs and agricultural practices block private ownership and industrial development. The phrase "usage of the land on a more permanent basis" suggests a rationale for the Chinese takeover of native land, a kind of manifest destiny. In fact the phrase comes from James Wong's remarks in the summer of 1987, after the Penan had blockaded a logging road in Wong's timber concession in Limbang. The road was built by Japan.

THE JAPANESE hold a unique place in East Asia, a place that few Americans learn about in school. In 1941 Japan was the only East Asian country besides Thailand that had avoided being conquered and colonized or, in the case of China, dominated and humiliated by the West. They had achieved this not by a passive "bend but do not break" policy like Thailand's but by retaining active control of their entire society. It was not by chance.

The Japanese had been well aware of European incursions into China, "the celestial empire," and by 1850 the spice trade of the "Far East" (the phrase recalls a Eurocentric point of view) had for two hundred years put Asian ports and islands under foreign control: the Portuguese in parts of India and Malaysia; the Dutch in Malaysia, Indonesia, and the Moluccas (the "spice islands"); the British in India, Burma, Malaysia, and in the Opium War in China; the French in Cambodia and Vietnam.

The Japanese, however, had kicked out their Christian missionaries in the seventeenth century; and except for a trading enclave at the port of Nagasaki, they had isolated themselves. Americans tend to see this as pulling back into a shell, but the more one considers European predations in East Asia from 1600 to 1941, the more prescient the Japanese policy seems. They knew they would lose in a confrontation, and they saw what was happening to other countries with open doors.

Finally, when Commodore Perry's "black ships"—a phrase every Japanese still instantly recognizes—sailed into Tokyo Bay in 1854, many Japanese knew they would have to adjust, and doubters were convinced over the next ten years, when in several engagements, the western military proved vastly superior. The cry of "Expel the barbarians" was tempered to "Rich country, strong military," to be achieved by learning barbarian skills. They bought time, negotiated treaties, and by 1868, with the Meiji restoration, they had made the internal changes that promised the kind of trading partner the United States demanded. While appearing sufficiently compliant to avoid confrontation, the Japanese had also retained control of their nation. This scenario is not obsolete.

In the 1870s, a new Japan had to figure out its relation to those European powers who had taken whatever they wanted, wherever they had found it. The Japanese sent delegations to France, Germany, England, and America to study laws, police systems, courts, constitutions and parliaments, armies, weapons, and means of production. Within fifteen years they were manufacturing their own warships and by 1905 had stunned the world by defeating the Russian navy in the Pacific.

The Japanese are capable of being proud. They saw no reason why an Asian nation, in response to European aggression—military and economic—had to be a colony, instead of a competitor. Just as James Brooke had trouble with the Chinese precisely because they were more British—indeed, the "Protestant work ethic" seems a weak reincarnation of Confucian values—our troubles with the Japanese often stem from their sameness: competitive, tough, proud, energetic, hardworking. Current American criticism rankles them partly because they think we should admire what they are doing; they have rebuilt as MacArthur directed, they have sped up their economy along industrial lines—and after all, they never asked us to slow down. Many Japanese perceive us as racist: we will accept competition from Europeans, but not from Asians. Also widespread is the belief that the atomic bomb was used on Japan but would not have been dropped on Germany. (It was, in fact, used as soon as developed, too late for the German war.) I believe they are profoundly wrong on this, underestimating man's ability to hate his brother, but the belief that the bombing was racist is widespread. The idea of a nationalistic economy, versus one based on individual efforts, is perhaps a profound difference between us, but in character we are much alike.

These similarities in character have shaped twentieth-century history. By 1900, Japan thought of itself not as a potential colony, but as a potential colonizer. From 1910, when it took Korea, to 1941, its ambitions to do what Britain had done—after all, it was Japan's own backyard—became more obvious; and not surprisingly, the colonial powers did not relish a new kid on the block. Japan, with no natural resources of its own except coal, would have loved to be in control of the oil fields at Brunei and Miri, for instance, but they were in the hands of Britain's Shell Oil. Japan's attempts to build navies and secure resource bases in Asia, crucial to a developing industrial nation, were systematically thwarted by Western powers. The nationalism, xenophobia, and racism of Japan in the 1930s grew partly from Western rejection of their "western style" ambitions. We would do well to remember that, as once again we rap their knuckles because of their success.

By 1941 the Japanese had developed a view of their role in history which included Asia for the Asians. America and Europe were preoccupied with Hitler; the time seemed ripe. The attack on Pearl Harbor was not really an attack on America, so much as an attack on America's capacity to protect Europe's Asian colonies and interests.

The Japanese role as liberator had a solid basis. Across Asia, people had indeed had enough of the whites, as Rajah Charles had predicted and postwar events would prove. In Sarawak, the small-time rubber planter William Wong had taken a British partner named Montgomery; when times turned tough, it did not take Wong long to think Montgomery racist, aggressive, insensitive to the Chinese family and to elders, and to see himself as the victim of British arrogance:

I feel inclined to write direct to the Rajah giving all the bullyings he gave me—there are quite a few dozens. But I always calm myself taking it as my fate and destiny to be brought up by Europeans and then bullied by Europeans too. [1937]

It is most curious through long, unhappy association with Mr Montgomery and feeling his bullying, overbearing aggressiveness and bad tempers, my mind is getting on to a point that I am getting to shun any European I see—a sort of dislike of looking or talking to them. [1938]

Then, two weeks later, directly to Montgomery:

> Look at your letter in answer to my request of 11th Aug. . . . How rough and unkind the wordings are, and what impressions they create in the children's minds. I am a man of 50 with much wider experiences in Home life, with many children to call me Father and am a Grand Father already. My head's bald and hairs gray and limbs weak, you should in your exalted position as an English Gentleman and our Rajah's Countryman, condescend to give some regards and respects to an elderly man like W Wong so to help him in Spiritual and Moral uplift in his life duties, and not treat him as Inferior Complex. We are all in sight of God same, except I was born an Asiatic and you an Englishman in a progressive age of the West. . . .

Even if the English were a burden, however, Asia was not necessarily looking forward to its liberators. The Japanese occupations of Korea, Manchuria, and China were unusually cruel, and word spread quickly. It was hard to know what to expect; people like William Wong could only hope for the best. Papa Wong on July 28, 1941: "War. The aggressor is now in Indo-China, and Papa am thinking whether it would not be a good idea to intensify planting of tapioca and sweet potatoes." An army will march on its stomach, and the family will have to be fed.

By September, the British, too busy with Hitler to consider defending Asian interests, had closed the oil fields in Borneo to prepare for demolition, to keep the oil out of Japanese hands. Europeans began to leave. At first, in 1939, William Wong had hoped the war might be confined to Germany and France, leaving the rubber trade intact; then he hoped Japan would stay out, and Pacific shipping continue. Then Pearl Harbor, December 7, 1941. The British torched the Brunei oil field. On December 9, Wong wrote to son Henry:

> You know now war is on. Steamers may or may not be running, and mails will be irregular. I do not know where Ko Ko James is: perhaps still in Singapore, perhaps in Kuching, perhaps reaching here tomorrow.
>
> I have got a new boat made using sail and paddle [anticipating no gas supply], and several nets made, all brand new. If no work then revert back to former times, singing and paddling along rivers,

fishing and cooking usefully as we did so usefully for our home before. . . .

I saw a very red glow in the sky last night lasting for hours, in the direction of Ulu Brunei.

No one thought the Japanese could sweep south so fast. Pearl Harbor on December 7, Malaysia mainland a few days later, Sarawak's Miri on December 21. New Year's Eve they occupied Singapore, and back in Sarawak, Limbang. British consuls in Kuching and Sandakan had not even evacuated women and children, thinking they had weeks to spare. They were all imprisoned. Agnes Keith's *Three Came Home* is a superbly fair and vivid account of a Japanese concentration camp in Borneo, and a pretty good movie. Son James had indeed been caught in Singapore, and was lucky, through his school, to be working with a Red Cross crew: "Though the Japanese rounded up thousands of Chinese in Singapore and shot most of them, our unit was not affected. After four months it was disbanded and I found myself living in a seminary dormitory with eighteen other young people from Sarawak and Sabah, who elected me their chairman. . . . Three months after that I was able to persuade the Japanese to send us home." The morning of January 1, when William Wong sat down to his typewriter, the horrors of Singapore were unknown. Although many of the women had been moved inland, Wong, and most Southeast Asian men, assumed that the Japanese were not after *them:*

> Japanese army invaded Limbang at 6 pm and took over without firing a single shot. We chit chat with them. They seem decent folks. They were well trained from childhood in this trade of military training. All short but strong and young and can run for miles without fatigue. This is the outcome of exercising and training daily.
>
> Well hope for the best. Everything is ordained by God. Things do not come just like this. Some things were wrong somewhere, and I think Japanese are coming to right same.

So for Wong, the occupation began peacefully enough, and offered the faintest promise of justice. The rubber trade had stopped, but he and his family muddled through the war on their gardening skills, supplying fruit to the occupying officers. By 1944, Wong's reputation as a gardener had brought him the possibility of a fat contract; the Japanese wanted him to estimate the cost of opening land and supplying food for ten thousand men. In a long and honest reply, he showed in detail why, in that particular climate and soil, the plans as envisioned were impractical.

By 1945 the occupation atmosphere had soured. The Japanese were losing, and naturally, as the Allies began to advance north from Australia, the Japanese were looking for scapegoats and opposition. Moreover, though William Wong probably was not aware of it, a very effective native resistance, led by Tom Harrisson and other Allied officers who had parachuted into the highlands, was giving the Japanese increasing trouble. Limbang had become a town of spies, counterspies, and informants. On January 5, 1945, William Wong records:

> Early at 6:30 a.m. Bakar the Arab just released from custody came and visited me for about fifteen minutes. We merely hear him talk. I suspect he is released to spy on others.

On May 13:

> Was told by Police that CPO wanted to see me at 8 a.m. on 15th. My mind has always been unhappy whenever called by CPO or other officials—for from experience in the past no good can come from any such interview.

On May 15:

> Went to Police Station on call. Waited for 3 hours, then data re my birth &c were taken. So now I must keep a record of who's coming up and visiting us and what for.

On June 4, 1945, the entry is in a different hand:

> Father arrested by Jap—written by Henry.

Undated:

> We went to Brunei and inquired.
> Australians landed at Brooketon June 10.

Eldest son James Wong continues in the third person:

> James was among members of the family at the wharf in Limbang when Pap was taken away, handcuffed, to Brunei. Papa called to him, "Never mind, James, look after your mother and your brother

and sisters." James was crying. "Yes, Papa," he replied, through his tears.

Papa Wong was taken to Brunei. Within two weeks the Australians landed in Brunei and Limbang, the Japanese surrendered, and word reached the waiting family that along with a number of Brunei dignitaries in the prison camp, William Wong had been beheaded by the Japanese just before liberation, and his body thrown into a mass grave.

SUDDENLY THE WAR was over, and even more suddenly his Papa was gone. The twenty-three-year-old James Wong, with his father's energy and ability, and charged as elder brother Ko Ko to "look after your mother and your brothers and sisters," was now the head of a large Chinese family. He could not afford to return to school and become a professor of agriculture, as he had hoped—who would feed the others?

So, in 1946, I started the Limbang Trading Company. Within a few months, I had a branch in Brunei and my business there developed promisingly. . . .

The trading company gave me a base from which to consider diversification, and I was more and more attracted to the timber which seemed to me Limbang's greatest resource. A great area of forest was virtually untouched in the upper reaches of the Limbang River valley. At that time Sarawak's timber industry was in its infancy, concentrated on the coastal plains. . . . It was widely assumed that hill forests could not be economically logged. I felt this assumption to be wrong, and in 1949 applied for, and got, a timber concession area in Upper Limbang. After many trials and tribulations—and, indeed, failures—I finally managed to extract hill timber economically by mechanical means [tractors].

This success had far reaching consequences. From Sarawak's point of view, it laid the foundations of a hill timber industry. As others began to emulate me, this helped to transform the economy of the state. Timber income became an important part of the state's revenue. From my own point of view, it led to timber becoming my mainstay. As my interests in the industry developed over the years, I transferred the trading business in Brunei, in 1958, to my brother Robert to manage. Timber, not only in Sarawak but elsewhere, became my main business.

The third Rajah ceded Sarawak to Britain in 1946, and the antidevel-opment stance of the Rajahs, much to the relief of the Chinese, was now history. Britain in turn ceded North Borneo to Malaysia in 1963; new opportunities arose. By now James Wong, as a prominent citizen of Limbang, had been on most important councils, had timber inter-ests in upper Limbang and just over the divide in the adjoining Tutoh and upper Baram, and when Sarawak joined Malaysia in 1963 he was Deputy Chief Minister of Sarawak. His father would have been proud.

THE JAPANESE failure to nurture East Asian trust, in wartime and in peace, may be a great diplomatic failure, comparable to the Ameri-can failure to support democratic principles in South America. Both countries had the opportunity for considerable sympathy and coopera-tion; both have been squandering those opportunities in domination and short-term profits.

After the war, the revived Japanese "miracle" economy moved south, and once again Japan misbehaved. Japanese government agencies, through loans, grants, and aid, and Japanese companies through de-velopment and joint ventures, gained a reputation for ruthless self-interest, including a willingness to work closely with ruling elites no matter what their political practices, and to participate in illegal schemes. In oil, timber, fishing, manufacturing, and even foreign aid, Japan became known for a kind of imperialism. Japanese activity re-sembled that of the United States at the same time in Nicaragua and Chile, for instance, and like the United States, Japan was trading im-mediate profits for the long-range goodwill that could give it leader-ship.

Take Indonesia as an example: Japan has been its largest inves-tor since 1969, making 32 percent of all foreign investments. In 1980, almost 50 percent of Indonesia's exports went to Japan. As in its trade with all of Southeast Asia, Japan has managed to import raw materi-als, and export finished and high tech materials. In 1980, 98 percent of Indonesia's exports were in raw materials, food, and minerals (the situation has improved with a ban of log exports and with sales of fin-ished plywood to Japan—a move Japan bitterly opposed); meanwhile, 85 percent of Japan's exports to Indonesia were industrial. The issue is fairly simple: many claim that instead of using aid and joint ventures to transfer technology, and to help poor countries become vigorous trading partners in the industrial world, Japan wants only to create economic colonies that supply cheap labor and raw materials. Japan's relations with Sarawak, according to Sim Kwang Yang, a Sarawak member of the opposition party, amount to "economic imperialism, pure and simple."

In 1974, anti-Japanese sentiment erupted in Indonesia. On January 15, during the visit to Jakarta by Japanese Prime Minister Kakuei Tanaka and his wife, students rioted, burning Japanese cars and motorcycles and pushing them into the river, attacking Japanese dealerships and shopping centers and even shops with Japanese goods. They marched on the Japanese embassy and the State oil company, which was involved in illegal price fixing with Japan, and toward the presidential palace. Prime Minister Tanaka and his wife were airlifted out by helicopter. Tanks and police rolled in, hundreds were wounded, eleven killed, thousands arrested.

The regional press was not entirely unsympathetic: while of course the students' methods were condemned, two Singapore papers noted that the demonstrations reflected fears of Japanese economic colonization in Southeast Asia, and the editorial in the *Straits Times* indicated that Asia wants to trade with Japan on a more equal basis, whereas at present, Japan does not train local managers, does not reinvest profits in the host country, and seeks only to expand its exports to the region. The *New Nation* mentioned the greed of Japanese investors. Even Tokyo's *Japan Times* called for Japan to reform its business ethics in Southeast Asia, and a professor from Kyoto University suggested that Japanese trading houses in Asia did indeed participate in corruption and unfair trade practices, and increased the disparity between rich and poor.

The Japan International Cooperation Agency (JICA) is one of three main instruments of Japanese overseas aid. This agency does not initiate projects; it receives requests from host countries. In 1986, it agreed to help Sarawak build a road into the forest in order to help doctors reach the rural poor.

That was certainly a laudable aim. The road, however, was in James Wong's timber concession in upper Limbang, near the Baram border, and Japan's largest trading house, C. Itoh, had the exclusive rights to Wong's timber. The road was capable of carrying logging trucks as well as doctors. It opened up remote virgin forest, in Penan territory, near Long Napir.

Seven miles from Long Napir, at an elevation of two thousand feet in hilly, moss-covered rainforest, is the place called by the Penan *lamin lajang,* or "the place for the cooking pot." In March 1987, Penan and Kelabit began to gather there, erecting the temporary stick and thatch shelters called *sulap.* Girls collected *isang* leaves, and older women wove them into roofing. A reporter from the Consumers Association of Penang (CAP) attended:

> The children were enjoying themselves, climbing vines, exploring the forests around and teasing their pet monkeys.
>
> By midafternoon, the gathering mist was getting thicker. The women stoked the fires to keep warm. Cooking pots bubbling with rice grains hung over them.
>
> By the next day, 31 more families of Penan were expected to arrive. Some of the children had gone to meet them in the forests to bring them here. At least 250 Kelabit and Penan men, women and children were gathered at the blockade.

The blockade was opposite the Stampin Timber camp of Wong's Limbang Trading Company, on the road funded by JICA. Along Saga, a Penan, told CAP:

> I walked a week to this blockade because I don't want my children, my wife and grandchildren to continue suffering from the destruction of our forests. We've lived in the forests for generations, before the British. Now the company has turned our water muddy, our food and plants and animals are gone, our graves damaged.

He explained about the graves:

> "For generations the Penan have lived without money but survived because the land is always there to give us what we need."
>
> Relating the story of how the manager of the Limbang Trading Ltd Company tried to give him a $100 note in "compensation" for the destruction of the graves of his parents and five other relatives he said: "I told him that our bodies, dead or alive, were not for sale and I pleaded with him that if they had so much money already to

please leave our land alone. But he just laughed and said the company had a license to work on Penan land . . . there is no such thing as Penan owning the land. Take this money or you will get nothing!" Said Along: "I still rejected the money."

Another Penan there, Rosylin Nyagong, said: "The company destroyed our forest, our food and rotan. We'll stay here until they listen to us. . . . Without our forest, we're all dead."

The blockades quickly spread, in March 1987, to Uma Bawang in the Baram, where Jok Jau's mature fruit trees had been bulldozed by loggers, and to a dozen other sites, including several belonging to Samling Timber. Our friend Jewin Lihan and his Penan at Long Bangan blockaded Baya Timber's road nearby. Jewin's wife Paya said: "We have done no wrong. They in turn destroyed our land. We want the authorities to help us, since we have no power." Long Bangan had blockaded in 1985, when the actions in the Tutoh were first organized, and again in 1986.

Tempers were rising. A Japanese team of liberal doctors (sent by JATAN, the environmental group in Tokyo) visited Long Bangan, where later we partied with our friend Richard and his relatives. The doctors confirmed the Penan complaints about the effect of logging on nutrition, and about harassment of natives in general. Indeed, while the doctors were there the manager of the neighboring timber camp, belonging to Baya Timber, appeared and demanded women for his men. Richard remembered the incident well. This happened several times.

International coverage of the many blockades increased, and in June the government seemed to respond to the embarrassing publicity. Several natives—and they were in fact the opposition—were flown to Kuala Lumpur to present their case directly to the Malaysian ministers. For the first time, native hopes soared. But fine words were not followed by action, and in the forests, both sides hardened. In many places, timber activities were at a halt. James Wong brought legal action against the Limbang blockade. On August 29 the police disassembled the blockade, but it was reerected in September. As usual, Wong was willing to speak to the press, and he made statements about the destructiveness of native shifting cultivation and his preference for "usage of the land on a more permanent basis." Wong was by this time Minister of Environment for Sarawak. SAM, as usual, took him on:

It is absurd that a timber tycoon owning 100,000 hectares of timber concession in Limbang and responsible for large scale forest destruction, should become Environment Minister. . . .

Recently, the Minister had claimed that logging did not damage the forest, that after five years there would be no difference between a logged forest and primary forest; and that there would be more trees and wildlife in the forest after logging.

This kind of statements has made him and the Sarawak Environment authorities the laughing stock of the scientific world.

Wong responded by saying that "logging is my bread and butter," which brought howls of derision from the Penan: what did the millionaire minister think the forest was to them?

At the end of October 1987, the Sarawak government struck back. Police dismantled all blockades and detained ninety-two people. Harrison Ngau was arrested and documents were seized at his SAM office in Marudi, under the Internal Security Act, "to defuse racial tension," the *Star* reported. Under the Security Act, one can be detained for sixty days without an order, and two years without trial with the approval of the Home Minister. Ironically, James Wong was detained under the same act in 1974 for sixty days, then sent to prison for thirteen months—all without charges or trial. Presumably he was suspected of conspiring to have Limbang secede from Malaysia and join neighboring Brunei.

News of the extraordinary native opposition of 1987, and of the government reaction, reached Japan, and a few people began to ask questions about that road in Limbang. One of those people was Yoichi Kuroda of JATAN.

Kuroda discovered that James Wong had formed a joint venture with C. Itoh, the world's largest trading company, in 1980, to work his concession at Long Napir, upper Limbang. In 1983 they planned a road from Long Medamit to Long Napir, about 28.5 kilometers, constructed it in 1984–85, and began logging. From Long Napir they planned to extend the road over to Long Seridan in the Baram; Long Napir was at the junction. The new road was also to be rented out to other concessions, who would pay C. Itoh and Limbang Trading a fee.

JATAN had just been set up when in April 1987 they received news of blockades on a new road at Long Napir; and through inside information in Tokyo, they learned that the road had received JICA funding. They were slipped some private documents. JICA had financed a 200 million yen loan to C. Itoh, matched by a 200 million yen loan from Japan's Import-Export Bank, which is one of the world's largest banks, uses public funds from the Japanese postal savings program, and as of 1993 issued no reports of its investments. Most logging bridges and roads are funded by this bank. "It is almost impossible to get information from the bank about its projects," says Kuroda. The two loans were guaranteed by the Dai-Ichi Kangyo Bank, which was the world's largest, and—no surprise—a member of C. Itoh's *keiretsu,* or conglomerate. Such close cooperation within a Japanese *keiretsu,* and between the *keiretsu* and the government, is typical of Japan, and of the way its business and politics have been coordinated in its dealings with Southeast Asia.

JATAN brought the case to the Diet (the Japanese parliament), since JICA funds are supposed to be spent for the welfare of the local people, and the local people of Sarawak seemed less than enthusiastic about this project. Kuroda wondered if this might not in fact be a commercial loan. JICA at first argued that their half of the road was the half outside Wong's concession; they also said that the natives were benefiting from their new access to medical care. The Diet grilled JICA and the Foreign Ministry for an hour at the end of July, and later that summer.

The resolution was typically Japanese. No laws were passed, no regulations or budgets changed, but several influential Diet members got on the phone to friends in the Foreign Ministry, a number of people in JICA were shifted to other offices, and in September C. Itoh announced it had paid the loan back to JICA. On October 5, 1987, C. Itoh closed its office in Limbang, sold its shares in Limbang Trading to Wong and to Thien, Sarawak's largest and richest timber concessionaire, and stopped their log imports from Wong. A few weeks later, the police moved on the blockades and made their arrests—although not until C. Itoh was gone, and Japan's hands were clean. It was a typically Japanese resolution because everything in the instance had been rescinded, while nothing in the system had been changed. Apologies, yes—but the companies and government were free to do the same thing again.

And so it seems that the minister in Sarawak, James Wong, and the ministers in Japan at JICA, and Wong as businessman and the traders at C. Itoh, were all able to use funds from the Japanese public to build a road through the forest of the Penan public. Yet the Japanese public did not know that it was paying for it, and the Sarawak public did not want it, and would not share in the profits. And the financing was arranged when a minister in the poor government asked the rich government for help, on his own timber concession, in the name of the people.

I first met James Wong at his Kuching office in the modern government tower outside of town, across the river, where the Rajahs used to live. All governments know good real estate. Outside the modern building, under the huge portico, liverymen lounged by parked Mercedes. Armed guards directed me to the information/security desk, where a handsome Malay officer called upstairs to verify my appointment with Wong. Up in the elevator, down the paneled hall to a spacious reception room. Yes, the attractive secretary replied, Mr. Wong is expecting me and will be off the phone in a minute, could I please have a seat. Here are some of his publications.

I am a little cool in the air-conditioned suite. Wide picture windows look across the farmed landscape, dotted with houses, to the city of Kuching, still picturesque in the white walls and red-tiled roofs of Chinatown.

James Wong is about six feet tall, and fills out his dark silk suit, yet is almost lost in a wood paneled office the size of a front yard. He is affable, open, and his English, of course, is excellent, thanks to the missionary schools and Papa Wong. I tell him I am researching a book on Sarawak and Japan, and would like to hear his side of the timber argument. He says he has nothing to hide, his files are open to me, that the logging is careful and selective—unlike our clear-cuts in the United States—and he wishes he had the time to take me to Limbang and show me himself how "the forest is regenerated, and there's more biomass and more game after the logging." I think he believes this, but an old Sarawakian friend of his told me he's sly as a fox. At any rate, the affability is charming. His open mouth has endeared him to environmentalists the world around. ("The forests grow back fast here; it rains too much, it ruins my golf game.") He goes on and on, sometimes surprising me: "No, of course this isn't a sustained yield. The situation is very bad. Sustained yield would be eight million (board feet) a year, the cut is now sixteen. We must come down to sustained yield, but the state government needs the income." What will the state do for revenue after the timber is gone in seven years? "We don't know. Maybe dams and water projects. Maybe not." I ask about native land rights. "There are no native land rights," he says, indicating it is all state trust, "or else the natives would destroy the forest with their slash and burn agriculture." I point out that native land rights seem to be specified in articles 5, 10, 15, and 18 of the Malaysian Constitution; he shrugs impatiently. Finally, I ask if the Japanese are investing in Sarawak, or only a market for his product. "They are a market only," he says curtly. "They are not investing a thing. Soon they will be finished and gone." He looks out the window, and I keep the silence.

He does not know that I have read his book on his father.

After the interview, he gives me copies of his numerous publications—replies to journalists and environmentalists. I like Wong, feeling that he has truly deceived himself about what he is doing. But I have a history of naiveté.

In the 1920s, in British North Borneo, Papa William Wong decided that gambling was bad for the Chinese, and bad for everyone. He was infuriated that the government would make money off gambling, and the opium trade. He wrote a letter to the government. The government's reply was to ask how, if gambling were abolished, it could replace the $25,000 a month revenue. Papa Wong's answer (emphases his), suggests a legacy his son, and Sarawak, might want to remember:

> A Government is looked upon as a Parent, a Guardian of the rights of the crowd or subjects, and that whatever it aims, it must as is her bounden duty to aim and strive for the good of the crowd, that whatever the majority say and common sense say is evil, the thing is EVIL and it should be stopped and not encouraged. That Government should watch the interests of the subjects, and foster virtues on the State, and it should not for the sake of monetary gain sell the virtue of the state. That there is no ENDS to justify the MEANS, when the people are suffering, that it should not enrich a FEW at the poverty of the MASS. IT IS A DUTY OF GOVERNMENT TO WATCH OUT FOR THE HEALTH, & GOOD OF THE PEOPLE OVER WHICH THEY GOVERN, and should she violate this SUPREME PRINCIPLE, she should clear out.

CHAPTER 9 🐾
BRUNO MANSER AND
THE PENAN

The road Japan funded on Wong's timber concession cut right through Penan territory in the Limbang district, and was clearly aimed across the border to the upper Tutoh, in the Baram. In the Tutoh area, at that time, was a small Swiss man, thirty-three years old, who had sought out the Penan and was living with them. He was perhaps the first European to learn the Penan language well, and to live—not just visit or observe—their way of life. This is remarkable because the nomadic Penan, the "wild men of Borneo," had fascinated Europeans for over two hundred years. In many ways like nature itself, they were admired, overlooked, misunderstood, abused. About three hundred nomadic Penan are left. This is the story of Europeans, Malay, and Chinese coming to them and to their forest.

THE RAJAH JAMES BROOKE and his friend Alfred Russel Wallace used to sit up evenings, in 1854 and after, at the palace in Kuching or at James's mountaintop retreat, a few hours away by canoe and foot. They took delight in arguing Wallace's new views on evolution, which he was developing concurrently with Darwin. James howled at the idea that his ancestors were orangutans.

Wallace had come to see the "little men." Just as the Galapagos and the voyage of the Beagle were inseparable from Darwin's intellectual growth, Borneo and the Malay archipelago shaped Wallace. Unlike Darwin, however, Wallace did not just sail through; he stayed eight years. He came to know the natives; and this, in turn, shaped his view of evolution and the descent of man.

In 1855, Kuching was lively. In the sixteen years since Brooke's arrival, the "very small town of brown huts and longhouses made of wood . . . sitting in brown squalor on the edge of mudflats," which he had described in 1839, had grown to include his attractive cottage-palace (The Grove) and several other British style bungalows on the landscaped grounds of the compound. Wallace and the Rajah looked out across the river to the thatch-roofed Malay houses and a growing Chinese bazaar with apartments above the shops. They could see hundreds of buildings through the tall jungle trees arched over the river, where birds hunted the mudflats and crocodiles lay in the sun. The casual, energetic and fun-loving Rajah set a good table and talked late into the night, often to visitors from Singapore. The Rajah's unique kingdom and hospitable palace were favorite stopping places for British travelers on their way to various colonies. Wallace, two years in the archipelago, told James he wanted "to see the Orang-utan, to study his habits and obtain good specimens."

The orangutan in Malay dialect is quite human (*orang utan* means "little man," as *orang tuan* is "headman" and *orang ulu* is "remote man," upriver, an aboriginal). But if the Malays spoke of the ape as a kind of human, Europeans were positively obsessed with the possibility. In his book on Borneo, Redmond O'Hanlon tells how in England, as early as 1792, Lord Monboddo "maintained that the orang-utan was a variety of *Homo sapiens* with a merely accidental speech impediment," and he "took his own pet ape out to supper parties dressed in a dinner-jacket to prove his point."

Wallace was fascinated with this leading candidate for a missing link; he shot seventeen. A few years later, in 1865–67, "the great Italian" as Tom Harrisson called him, Odoardo Beccari, came through Sarawak with his gun:

> Looking intently, I at last made out something like red hair amidst the dense foliage. There could no longer be any doubt—it was an orang recumbent on its nest. The creature was evidently aware that it had been discovered, and yet it showed no fear, nor did it attempt to fly. On the contrary, it got up and looked down at us, and then descended lower amidst the branches, as if it wished to get a better view of us, holding on to the ropes of a creeper which hung from a branch on which it was first squatting. When I moved to take aim with my gun, it hauled itself up again, pushing forward its head, to look at me as it held on to the branches above with its hand. It was in this position when I fired. . . . I caught sight of a second orang on another nest. Although I couldn't see it well, I fired. . . . I perceived

something reddish moving on the top of a big tree. I fired at once almost at random. . . . As I was reloading, a second suddenly appeared, . . . [then] the huge beast turned, and it fell dead to a bullet in the chest. I should particularly have liked the skin, but I had to abandon both it and the skeleton and content myself with the head alone.

It was quite dark when we reached the camp, loaded with orangoutangs, drenched to the skin. . . . All told, I had got either the entire skeletons or portions of twenty-four individuals. Later, Azton brought to me several other heads from the same district. But with all this I came away without having been able to solve the doubts I had regarding the species or races of orang-outang.

"Hoorah for systematic collecting!" exclaimed Harrisson, quoting this passage in 1938. A few years later Harrisson returned during the Second World War, to direct the resistance from the Bario highlands. He stayed on as curator of the Sarawak museum.

The orangutan—an ape-man both theoretically valuable and physically expendable—wandered into European lore during the great romance of nature, 1750–1850, when Rousseau's "natural man" was a widespread ideal. Living rudely yet happily in a state of nature presumed the opposite of a state of civilization, the natural man was thought to be simple, innocent, direct, unspoiled. Holding to this fantasy, Europeans throughout the colonized world (on which the sun never set) overlooked the tribalism and complex socialization of natives, called them "primitives," and alternately envied them as free men and enslaved them as beasts.

Deep in the jungles of Borneo, which until the logging roads of the 1980s and 1990s was one of the wildest places on earth, lived—also—*Homo sapiens*: nomadic, thinly clad forest hunters with blowpipes. Given the drift of European thought, it is not surprising that the orangutans sometimes got confused with the Penan. Before he left England for Borneo, in 1838, James Brooke said he had heard of men "little better than monkeys, who live in trees, eat without cooking, are hunted by the other tribes, and would seem to exist in the lowest conceivable grade of humanity." And he wrote in Rousseauan manner that he "wished to see man in the rudest state of nature." By the late nineteenth century the desire to push orangutans forward toward men and nomadic men back, toward orangutans, in order to bridge the great man-nature gap imagined in Europe, had led to accounts of the "wild men in the interior of Borneo" living "absolutely in a state of nature, who neither cultivate the ground nor live in huts; who neither eat rice nor salt, and who do not associate with each other, but rove about some woods, like wild beasts; the sexes meet in the jungle, or the man carries away the woman. . . ."

The "singles bar" fantasy at the end is distinctly European, as is the notion of isolated individuals roaming alone. Even as recently as 1969, the "brown nomadic hunters, the Punan and Ukit" of Borneo, were identified with "the paleolithic virtues and vices of keen eyesight, alert observation, and incapacity for what neolithic man calls work."

Brooke's fantasies of a "natural man" live on to the present day. In November 1990, I was sitting at dinner with Chinese in Kuching, listening to one man tell of meeting (after flying into Mulu Park by company helicopter) a Penan man who could not tell how many children he had. This was interpreted not as a failure of communication or of numbers, but as evidence of a social structure so "wild" that the Penan man had no idea who or how many were his children (horror to the Chinese, and sheer nonsense). The entire incident was adduced as proof that their condition is so "degraded," so "barely human," that it is a favor to take the Penan out of the jungle, or, more to the point, to take the jungle away from the Penan. The Chinese gentleman was trying to show me why logging benefits natives. Having taught Native American history, I felt the weight of centuries as I heard, yet again, a man justifying colonial exploitation by means of racist attacks on those born with the gold, the silver, the buffalo, the wood.

O'Hanlon, who wrote a book on Conrad and Darwin as well as *Into the Heart of Borneo,* so funny and humane it could be Irish, notes that Darwin, like my Chinese friend in Kuching, spent very little time upriver. O'Hanlon, lying on a Sarawak riverbank considering his Iban guide Leon and wondering how anyone could think these people stupid or insensitive, recalls that among the evolutionary thinkers, only Wallace spent time with native guides:

Helped at every turn by Leon's ancestors and related peoples in his eight years of travel, often by native *prau,* from one island to another in the Malay Archipelago, he [Wallace] had come to conclude that "The more I see of uncivilised people, the better I think of human nature, and the essential differences between civilised and savage men disappear." He developed his concept of Latent Development—all the races of *Homo sapiens* had evolved a much bigger brain than they actually needed, at the same time. They just used different parts of its capacity in different ways.

Wallace found the interior natives beautiful, energetic, resourceful and clever, "lively, talkative . . . truthful and honest to a remarkable degree. . . . Crimes of violence (other than head-hunting) are almost unknown." James Brooke shared, and probably helped shape, this opinion.

So the natives in Borneo, and evening debates in Kuching, may have helped to give Europe its least racist theory of evolution, and Sarawak its least racist Rajahs, for on one point Wallace and all Brookes agreed: the natives were extraordinarily fine human beings.

The origins of the Bornean natives are obscure. The oldest evidence of *Homo sapiens* in all of Asia is found in north Borneo, in the Niah caves and elsewhere, Stone Age remains forty thousand years old. Who these people were we do not know. The first Malay migration down from southwestern China was around ten thousand years ago, and the pockets of aboriginals (*orang-asli*) left on the mainland are probably descended from those people. They were supplanted, however, by "proto-Malays" about four thousand years ago, who were probably the ancestors of the Malays and natives of present-day Borneo. In the broadest terms, we can say that the people of Borneo—native, Malay, and Chinese—are one race, Mongolian, originally out of central Asia; that the natives and Malays are part of the great Malay migrations that repeatedly overran the peninsula, Java, and the islands thousands of years ago; and that the Chinese are much more recently arrived from the mainland. Across Southeast Asia, most of the people are Mongolian: pockets of Negroid natives remain in the Andaman-Nicobar Islands, Papua New Guinea, and Australia, and empires and emigrations have brought some Aryans and Dravidians from India to Malaysia, and many Caucasians to Australia and New Zealand. Still, Malaysia and especially Borneo are thoroughly Asian. It can be hard to tell, at times, whether you are looking at a native of Borneo, a Malaysian, a Chinese, or an American Indian. Bornean native groups migrate, and easily mix, merge, and split. "Malays" are those who became Muslim, and a ruling class, about five hundred years ago. Intermarriage between all groups has for centuries blurred the taxonomists' best attempts to keep men and women apart.

Languages have evolved to become substantially different and mutually unintelligible, so that language is one hard division; another is geography. The effective identity is by place, longhouse by longhouse, and by tongue. In the 1980 census, the Iban, that vigorous native group migrating north from Kalimantan over the last two centuries, could be seen to dominate Sarawak; the native groups up the Baram are minorities. Major native groups of Sarawak included the Iban (368,000), Bidayuh (104,000), Kenyah (15,600), Kayan (13,400), Kedayan (10,700), Murut (9,500), and Penan (5,600, probably undercounted). So the tribes represented up the Baram—the Kayans, Kenyahs, and Penan—all together amount to about one-seventh of the Iban. Clearly the issue of minority rights could arise, even within the context of a native-controlled government. The natives are about 45 percent of the Sarawak population, the Chinese 30 percent, and the Malay 20 percent.

The Penan, then, are a remnant of an ancient Asian invasion—the remnant that has stayed deep inside, away from the rivers, one of the last nomadic tribes on earth. Perhaps 300 of them (500 in 1989, 700 in 1987) still roam free in the jungles, eating fruit and sago palm and hunting animals with blowpipes, while 7,000 live in scattered settlements in upper stretches of Sarawak rivers, combining farming and hunting in shrinking habitat. Most of the semi-settled Penan have been forced into longhouses and rice-growing within the last twenty years. These are the people all the other natives know, with some awe and respect, as the ones at home in the forest. As nomads, they do not build longhouses. They were never headhunters. They are shy and secretive, coming out occasionally to remote longhouses to trade. They know you are inside the jungle, but you do not know they are present unless they choose to reveal themselves. Even among the other natives, who consider the Penan inferior socially and politically, who have sometimes treated them as property and slaves, the elusiveness of the nomads creates a certain mystery. They move every few weeks or months, finding clean camps and fresh foraging and game. They are legendary: tough, sweet, kind, wild—and scarce.

These are the people—the wild men of the forest, the missing links, the "little men" and noble savages—that Bruno Manser, in 1984, left

Switzerland to find. "And I can assure you," he said to us years later, in Tokyo, "that I have found these nomadic groups."

WE HAD HEARD of Manser before leaving the United States, but had little idea who or what he was. Then in August 1989, in Taman Negara Park in mainland Malaysia, a young woman wearing a white blouse and red skirt walked out of the jungle and into the river. She was blonde and blue-eyed. As her skirt billowed up, she sank to her neck with a sigh, closed her eyes, and let the current carry the heat away.

Hannah Olesan, twenty-two, from Denmark, was working on her master's degree on shifting agriculture in Borneo. She spoke quietly of primary forests, of forest farms and of forests gone, of wild pigs and orangutans in Kalimantan. Then she told of happening upon a native blockade of logging in Sarawak. She had been walking alone up the new dirt road. A logging truck stopped. As she climbed in, the driver said in English, "Do you know Bruno Manser?"

"That's what they always ask," she said to me with a smile and twinkling eyes. "They will ask you."

"Do you know him?"

"No. He is hiding with the Penan. They move him around in the jungle. I saw them at the logging blockade—five or ten natives behind wooden poles, holding blowpipes and knives. The police kept me away. Everyone said Bruno was somewhere near, in the jungle, watching. That's how the pictures got to the *Straits Times* in Singapore."

Bruno Manser was born in 1954 in Basel, Switzerland, one of six children in the German-speaking family of a gardener. At eighteen, after high school, he disappointed his parents by skipping the university and moving to a farming village. There, living alone, he worked as a shepherd and fisherman, grew his own food, and sewed his clothing. In his words, he "made practical apprenticeship in all fields of agriculture and handicraft, trying to get the base of economic self-sufficience and independence, spending most of the time in the Swiss alps. Six years as cheesemaker with cows, six years with sheeps." Small, wiry, impish—it is not hard to picture him beside his cows in Heidi's alpine meadows. By 1982, he was also interested in spelunking, and spent much of his time in caves deep beneath the Alps.

His mountain retreat, however, was not enough. The next phase of his life he told vividly to Wade Davis, rainforest expert, ethnobotanist, and writer, when Manser won *Outside* magazine's Outsider of the Year award in 1991:

> "As a child, I collected leaves and feathers, and at night lay in my room imagining that it had become a jungle. I wanted to live with a people of nature, to discover their origins, to become aware of their religion and life, to know these things." In a library he came upon a single black-and-white photograph of a Penan hunter with a caption that read simply, "A hunter-gatherer in the forest of Borneo." The book offered no other information. Manser dug further, eventually discovering an obscure report that described the Penan's homeland: lush forest and soaring mountains, dissected by crystalline rivers and the world's most extensive network of caves.
>
> Intrigued, Manser travelled to Malaysia in the winter of 1984, became conversant in Malay, and accepted an invitation to join a British caving expedition headed for Gunung Mulu, a national park that encompasses the heart of Penan territory in Sarawak [up the Tutoh]. . . . Manser and the British cavers traveled deep into the park, where they explored the caves for two weeks. The British then departed, but Manser had been told of a group of nomadic Penan living beyond the southern boundary of the park in a region called the Ubong, and so he pushed on, alone. For several days — most, Manser says, without food or water — he struggled through a dense jungle that seemed to mock everything he had learned in Switzerland about nature. On the ninth day, exhausted but reluctant to turn back, he climbed a tree and saw, across a valley, the white plume of a cooking fire.
>
> It was nearly dark by the time he came upon the footprints in the mud by the river. "I knew they would be afraid," he says, "so I made camp. The next morning I let the sun come up. Then I heard two voices, a man and a woman's. For two minutes nothing happened. I held up my hand in greeting. The woman fled. But the man came to me. He spoke a few words of Malay. We touched hands and he drew his fingers back across his breast." Manser followed the Penan man up the slope to the encampment.

Without special provision, the tourist's entry visa for Sarawak is good for three weeks. By December 31, 1984, Bruno Manser had overstayed his visa. At first, this was of no great concern to the authorities, and certainly not to himself. He was in settlements or deep inside the forest, up the Tutoh a few days' travel above Long Bangan, near Long Seridan, stripped to his shorts, toughening his bare feet, learning the blowpipe and the Penan language, beginning his valuable journals of Penan vocabulary, customs, and drawings, most of which are still in the hands of the Malaysian police, who have refused to release them. Wade Davis writes:

> Fearful of the heat of the sun, ignorant of the seas, insulated by the branches of the canopy, the Penan live in a cognitive and spiritual world based entirely on the forest. Distance and time are measured not in hours or miles but in the quality of the experience itself. A hunting trip, if successful, is considered short, though a Westerner might measure it in weeks. An arduous journey is one that exposes a Penan to the sun.

Manser, as much as possible, became a Penan, and they accepted him. Pictures from the jungle years show him in his bowl haircut, in shorts, with Penan bracelets on wrists and legs. Often he is seated on a tree limb, blowpipe nearby, Penan basket, notebook, and pet monkey in his lap. Sometimes, he is playing the flute. The puckish smile is actually Penan as well as Swiss; only the wire-rimmed glasses give him away.

In the 1980s, the Penan were distributed mainly across the upper Tutoh and upper Baram, and over the divide into the headwaters of the Rajang. A high percentage of the nomads were in the upper Tutoh, especially along the Magoh tributary (a few were also up the Silat and Tinjar, tributaries of the Baram). Like the grizzly bear in North America, Penan had been driven to the most remote, inaccessible habitats.

The nomadic Penan move from base camps, which may last a year, to temporary camps near food sources, and to travel camps—bivouacs—over a large territory. In all camps, nomadic Penan build light, thatched shelters of bamboo or softwood, one per family. Usually the huts are up on stilts, above the damp ground, above the leeches. The floors are split bamboo, well ventilated and springy, quite comfortable for sleeping. The roof is thatched leaves. Nomads do not build hardwood longhouses. A group of nomads might typically number thirty; those semi-settled in longhouses, from fifty to two hundred or so.

The rhythm of Penan life is determined by the maturing of wild sago palms, from which they pound a starchy paste, by jungle fruits, which mature somewhat unpredictably because there are no seasons, and by animal migrations. Besides the sago palm, their staple food, they harvest *lekak,* an edible palm bud, and various fruits, ferns, vegetables, and roots. Wade Davis, who has studied twenty tribes of the Amazon and South America, says, "the knowledge of the forest by the Penan surpasses all of them. It's unbelievable. . . . They recognize more than a hundred fruiting trees and at least fifty medicinal plants."

Sago palms grow in clumps, several trunks springing from one mass of roots. "If there are many trunks," says a Penan, "we will get one or two. We thin it out so it will thrive. If there is a lot of sago, we will harvest some, and will leave some. We don't like to kill it all off, in case one day there is nothing for us to eat. This is really our way of life. . . . If we harvest the [sago at] Ula Jek first and finish the *nangah* [mature sago] there, we *molong* [put a mark and preserve for future use] the *uvud* [young sago]. . . . After two or three years, mature sago will grow out of the young sago that we have preserved." Scattered across the steep, intricate ridges and valleys of their district, many wild sago clumps will be known and claimed (marked) by a single group. They know when it is time, in two or three years, to return to a certain clump. And so with many other types of edible plants; in all of their fruit and vegetable harvesting, they are careful to preserve a sustained yield.

The other rhythm is hunting, which they love. The small, bearded pig of Borneo is their favorite food, and in virgin forest, pig supplies most of their protein. The pig is taken with the poison dart from a blowpipe (rarely), dogs and spears or, increasingly, by shotgun. They also hunt, by blowpipe, deer, monkey, gibbon, civet, porcupine, squirrel, and just about anything else. The rhinos are gone, though fresh tracks were seen in the highlands in 1984. The pigs travel in bands, foraging their favorite roots, fruits, and nuts. Fruiting can be unpredictable, but the Penan know exactly where the fruit trees are, and live for the pig migrations to come their way. In one heart-breaking story of a few years ago, the small, clear tributary suddenly went brown, and the hunters jumped to their weapons—pigs must be crossing upstream, a lot of them. No, it turned out to be big, and yellow. A bulldozer.

By all accounts, the Penan and the longhouses can live well in virgin forest. One ethnologist in Belaga in 1986 reported that Penan hunts lasted from sunup to early afternoon, covered six or eight miles, and had a 90 percent success rate. A century before, up the Rajang River in longhouses in 1865, Alfred Russel Wallace reported, "The people produce far more food than they consume, and exchange the surplus for gongs and brass cannon, ancient jars, and gold and silver ornaments, which constitute their wealth. On the whole, they appear very free from disease. . . ." Eric Hansen, who walked back and forth across the most remote Penan territories in 1982, says, "The only Penan I have seen who were not in superb physical condition were from Sungai Ubong on the ulu Tutoh. . . . their traditional hunting grounds were squeezed between Mulu National Park and a huge timber concession." This is the Penan settlement the Chinese gentleman had visited by helicopter.

The Penan have for centuries traded with nearby longhouses, which in turn traded with Chinese boat peddlers working out of settlements downstream. From the forest they harvest camphor, wild rubber, *damar* (a resin), *gaharu* (the incense wood which boomed in the 1970s), bezoar stones (monkey gallstones valuable to Chinese medicine), and rattan, which they use to make the famous Penan baskets and mats, for their own use and for sale. They trade at longhouses for salt, cloth, tobacco, cooking utensils, radios, tape players, batteries.

Their personal style, and the atmosphere in a camp, are difficult to describe, and perhaps to believe. Mild and sweet in manner, curious and active, to us at least they are utterly charming. All food being shared, their governance democratic, their manner open, it is not surprising that generosity is a primary virtue. Davis reported of Manser that in six years "he never saw Penan quarrel. Only once did he see a hungry child neglect to share food." Manser said: "There was a boy in that first Penan group who caught seven fish. I remember watching the headman give three to each family and then carefully slice the remaining fish in two. That is the Penan. You will never find one with a full stomach and another who is hungry."

Readers from my own culture may feel that such a paradise must be lost. Alas, that is the story we are writing across the face of the earth. Bruno's first years with the Penan, 1984–85, were the years the logging moved up the Tutoh. In 1975, the annual cut in Sarawak had been 2.5 million cubic meters; by 1985 it was 12 million. By 1990, it was 18 million. In 1985, as Bruno was getting to know his new friends, who had probably been in the jungle for four thousand years, over an acre of forest was being cut each minute. The logging raced, very much like a wildfire, up from Marudi, up the Tutoh, past the park, into the district where Bruno was living. It happened so quickly.

An announcement in fine print would have appeared in the *Sarawak Gazette,* in Kuching, months, maybe years before. It would have given the natives six weeks to claim customary rights to tilled acres in a certain district. Nomads would have had no rights anyway, even if they had happened to see the newspaper (no mail service above Marudi), find someone who could read it, and compose a written reply within six weeks. The land now belonged to the government, and the government had seen fit to lease it out to a timber concession. Never mind that since the drifting apart of the continents, the rise of mammals, the first migrations from central Asia—that in all human history not a single government or its agents had ever set foot in most of this forest. Not the Malay sultans, the white Rajahs, the Japanese in World War II, the British resistance (who chased the Japanese up one stream to Limbang), the colonial regime, or modern ministers of Malaysia and Sarawak. Government had never left the rivers and the park. Suddenly the forest was declared the property of those who had never seen it, who did not know it, who did not love it. But they were powerful people, and they intended to be rich.

For the Penan, the first warning was the sound of a helicopter, just before Manser arrived. "The Penan expected to talk about their situation. Instead, they were stunned as an anthropologist asked to measure their skulls." The "little men" were not amused. The government survey party did its work, said the land belonged to the state and would be logged.

After four attempts in the fall of 1985, a few Penan leaders were finally able to bring representatives of Tutoh nomadic bands and settlements together in a meeting. To the Penan, such a political gathering was not an accustomed activity; you might imagine the level of fear, frustration, misinformation, and ignorance. Manser, too, knew nothing. As secretary to the group, he helped send a petition to the government to protect five hundred square miles near the park as a nomadic homeland. "I was so naive," he recalled. "It was such a small area of land. You could walk across it in three or four days. I really expected the government to set it aside." The petition letter was never answered.

Still hoping for quick changes, he prepared in 1985 a report on Sarawak logging and sent it to twenty papers and magazines in Malaysia, Europe, and the United States. Nothing happened, though of course some seeds had been sown.

Other native tribes, settled tribes with more political savvy, had already blockaded some logging roads before Manser arrived in Sarawak. He urged the Penan, traditionally shy and nonviolent, to do the same.

In the spring of 1986, as told by Wade Davis:

> Manser and a group of 25 Penan stood in front of an oncoming bulldozer that was breaking ground for a bridge into the forest. The driver retreated, but the next morning 30 bulldozers appeared at the roadhead, backed by police and logging company officials. The Penan fled. Malaysian officials had trouble believing that the savages of the forest could devise such a strategy of resistance. They blamed instead a foreign agent of influence and ordered his arrest. Manser became a marked man.

From 1986, when the police took official notice of him, until 1990, the story could be heard essentially unchanged in any upriver cafe, any airport police office, the Sarawak State Government offices in Kuching, the American Embassy in Kuala Lumpur, or in Friends of the Earth offices around the world. Bruno Manser, a Swiss, is hiding with the Penan, Malaysia has a price on his head, and he is organizing opposition to the timber trade.

It was a story Europeans wanted to hear, and also a story Malaysia wanted to tell, since it diverted attention from the new SAM office in Marudi, staffed by Sarawak citizens, natives from the longhouses who thought the timber trade was rotten to the core. The Manser story focused all attention on the romance of a white savior.

After the spring of 1986, Manser could not safely leave the forest and come in to the settlements. The companies and government were said to offer $35,000 for his capture, and the nomadic Penan kept him for three years, moving about in the forest, passing him from band to band. The two hundred armed soldiers sent after him by Malaysia could not find the camps. Manser began to learn the meaning of the stick signs left by Penan for each other at every jungle trail or event, who was going where, why, when, what's up. Two pieces of wood across the completed sign meant that all people of the forest were of one heart.

DIE JUNGEN PUNAN-MÄNNER SIND MEISTENS WUNDER-
SCHÖNE MENSCHEN MIT RABENSCHWARZEM HAAR,
DUNKLEN AUGEN, SEHNIGEN KÖRPERN UND
SEIDENFARBIGER HAUT. - DER TRADITIONELLE
SCHMUCK "CELUNGAN", ARM- UND BEINREIFE AUS
SCHWARZGEFÄRBTEM ROTANG, IST IM VER-
SCHWINDEN BEGRIFFEN, SOWIE ELFENBEIN-
ARMREIFE. SIE WERDEN HEUTE DURCH EI-
NEN GUMMIRING ODER KUNSTSTOFFREIF ER-
SETZT.
WELCH EINE FREUDE AUCH, DEN
HERANWACHSENDEN PUNAN-
KINDERN BEIM SPIELEN ZU-
ZUSCHAUEN. SIE NEHMEN
AM STEILHANG DEN
LEHM, SETZEN SICH AUF

Pages from Bruno Manser's journals.

"In all my years among the Penan," Manser says, "I never saw a sign that did not bear this simple message." Thom Henley of Vancouver, Canada, tells of an old Penan placing one rock in the middle of a circle of rocks, threatening sticks outside the circle, to show how the Penan sheltered Manser from the police.

Manser was captured twice, and twice escaped. The first time, by chance, he ran into an off-duty policeman who recognized him; he ran away, shots whizzing over his head. A second time he was betrayed by a reporter who came to a Penan meeting by helicopter, left, and then the police arrived in the same helicopter. He was taken down the Tutoh by boat; on the way, he had time to review his situation, and in a rapids he jumped out. Shots again were fired; he swam to shore and disappeared into the jungle. That is how the police came to possess his notebooks. In both cases, he thinks they were probably not shooting to kill, but the doubt was sobering.

In April 1987, the Baram erupted with blockades and logging oppo-

sition. During the summer months, fifteen timber operations were shut down by various tribes at over twenty-three sites in the Baram and Limbang, and four bridges were burned. It was close to an undeclared war; the government passed new laws against obstructing timber operations, stepped up harassments and arrests, and went after Harrison Ngau and Bruno Manser. Manser was a rajah, they said, carried by the Penan on a bamboo throne; a Zionist (Malaysia is Muslim); a communist (armed communist insurgents were still operating at the Thai border on the mainland). The Sarawak newspapers loved it. The *Star*, April 30, 1987, reported:

I'M SECRETARY TO THE PENANS: MANSER

Kuching, Wed. — *The Star* has received a letter supposedly from Swiss fugitive Bruno Manser in which he says he is hiding in the Sarawak jungle to act as secretary to the Penans.

The writer says the land of the nomadic tribe is being destroyed by timber operators and that the State Government and logging companies are not looking into the Penans' demands.

He alleged that one timber company had already destroyed half of the forest area in Magoh where the Penans collected their rattan to make baskets, mats and other items. He says good quality rattan is hard to find in other parts of the State.

He adds that fruit trees, sago palms and tacem, from which the Penan get poison for their darts, are being felled by loggers. . . . Manser, 33, has stayed on illegally since December 1984. He is said to be carrying out a study on the Penans. . . .

The State Government has, however, said that it would not recognise the study because it was being carried out illegally.

A government spokesman said that even before Manser entered the jungle, several government agencies had conducted studies and research on the Penans and had drawn up plans to improve their standard of living.

He said the Penans have been accepting changes slowly and interference from foreigners like Manser had hindered the government's efforts.

For the past few weeks, the Penans have been blockading timber camps, bringing logging activities to a standstill. The State Government believes that the action was instigated by Manser.

By 1988 the international community was responding; the European Parliament passed a resolution to suspend Sarawak timber imports, and debated banning them altogether; some Australian docks refused to unload Malaysian logs; criticism of the tropical timber trade mounted in Japan. Meanwhile, Manser, who had probably come to know the

Penan better than any European since interest in "the wild man of Borneo" began, continued his unauthorized study. The government pointed out that he was not a trained anthropologist and therefore his findings were worthless. Manser, however, seems to have found a superior way of measuring heads: Henley recalls being in a longhouse, listening to a tape that was circulating among the settled Penan. "Bruno was singing, in Penan, songs he had made up about the government. The people were convulsed with laughter. One of them explained to me, 'He understands our language. He makes very good jokes.'"

The opposition continued. September 1989 saw the most massive blockades to date, with four thousand natives participating in twelve roadblocks. International attention had been gained. Yet, not only had the logging continued, moving up the Tutoh past Manser's home into the Magoh, and up the middle Baram past Long San, it had increased. Day and night, the timber was coming out faster than ever, as if the companies and government were desperate. In five years, the annual cut had gone from 12 million to 18 million cubic meters. By 1990, it was clear that SAM and Manser and the natives scattered across the forest had succeeded in organizing themselves, had succeeded in staging protests and drawing international attention—and yet they were losing, losing badly. Only a few years were left for the Penan before a whole way of life would be gone; for the longhouses, a few more years, five to ten, remained. In the spring of 1990 Manser knew that what could be done from inside Sarawak had been done. Only world pressure, largely on Japan, could stop the cutting. He escaped from Sarawak—how, where, is a secret. But after five years of hiding deep inside, in March 1990, he came out. "I would have stayed," he said, "if the Penan could have been left alone."

ON THE AFTERNOON of June 6, 1990, Bruno Manser was out of the rainforest, sitting on the eighteenth floor of a building in downtown Tokyo, staring out the window: across the street, a platform of five window-washers hung halfway down a wall of glass and steel. They looked like ants, a cliché that after six years in Borneo must have had a force for Bruno that we cannot imagine. To their left, painted on a fifty-foot-high billboard, was a huge, nearly naked Chinese woman with a semi-automatic assault rifle—a movie advertisement. Behind her, the blue Matsushita Electric building, the black Sony tower, and the department stores of the Ginza. "I went to Sarawak," he was saying in a soft and careful voice with a slight German accent, "to join the life of an indigenous people who still live independent, having their own economy, somehow in harmony with nature." He looked at his audience. "They have existed."

Bruno Manser seems small in baggy trousers and a rough cotton, open-necked shirt, about five feet nine inches tall and weighing 150 pounds, but pictures from the jungle show an adventurer's muscles on a lean frame trimmed by malaria. He wears rimless glasses and has an instantly engaging smile, at once witty, ironic, and compassionate— Ben Kingsley playing Gandhi. That afternoon, his gaze roamed the room and the wry smile was directed at the dark wall panels of tropical timber, as he spoke:

> The settlements are dirty. They will throw just anything out of the house. But when they are in the jungle, the dish is a leaf and they throw the leaf away, and when their hut will get greasy from the wild boar fat or black from the charcoal also they will just leave for a new place—they will be all new and clean—and that's how they can survive.

Hosted by Yoichi Kuroda of JATAN and the Friends of the Earth office in Tokyo, Manser was meeting Japanese government and industry leaders, holding vigils on the sidewalk in front of Mitsubishi, running off mimeos in the tiny JATAN office in crowded Shibuya, downtown Tokyo, trying to find his footing on the treacherous path of international environmentalism.

Bruno Manser, sui generis, Swiss shepherd, intellectual, spelunker, botanist, artist, is possibly the first man, and maybe the last, to stand with one foot in the stone age, or at least in a European Romantic's love of it, and the other foot in the offices of environmental science: a world of timber exports, hectares per minute logged, carbon dioxide emissions, German analyses of satellite photographs showing 4 percent of the Philippine forest left.

His new allies are hardly Penan: bright, young one-worlders in tiny, messy offices which in 1960 would have churned out civil rights mimeos and in 1970 anti-Vietnam or pro-consumer Xeroxes and now have the latest fax on whales. They and Bruno are friends, but they prefer law school to leeches. Yoichi Kuroda of JATTAN and Thom Henley of Endangered Peoples Project are handsome and articulate, wear their suits comfortably and answer stupid questions with courtesy, smart questions with facts. Henley is impressive in Washington and Tokyo and has visited Sarawak four times. Henley and Kuroda, however, would not prefer to remain with the Penan.

Strange that a man so independent, who wanted to get away to a life in nature, should wind up at the middle of so much. Some of Manser's activities would later be seen as competitive to SAM, while some of his popularity in Europe offends even activist Malaysians because it seems racist. Randy Hayes of Rainforest Action Network sums it up well:

> Bruno Manser's story evokes the notion of the Tarzan syndrome. No one cared about the Penan until a white man came to the scene; they were considered little brown men who needed guidance. While this is not Bruno Manser's fault, he plays into it unwittingly. It's a classic European fantasy—a Lawrence of Arabia. . . . Manser's impact lies in the fact that he brought attention to the issue. But there's a danger in letting the messenger steal the limelight.

Even opposition Sarawakians and Malaysians can easily tire of hearing about the Rajah Brookes, and Manser. The white men, of course, were the ones who once thought the Penan were apes; who loved the orangutans to death; who brought the British land laws that now serve the government so well; who brought the helicopters, the bulldozers. And now, as international economies replace race as locations of power, the Japanese join the Europeans in the first world: these new Asian white men build the dozers, run the banks, buy and burn the logs. Being saved by someone else's Tarzan—that rankles.

There's another danger, besides the attention to the foreign adventurer: the Penan themselves, as nomads, play into the hands of European romanticism; also, as nomads, they divert attention from the issue of longhouse land rights, the overriding issue for most of the population and most of the territory. Therefore, the Penan issue is a double-edged sword: it attracts international attention, but makes an easy target for government defenses of timber. Do you expect to keep them in a stone age forever? Just 700 . . . 500 . . . 300 of them left? Over and over, Chinese in Kuching accused me of wanting to save the nomads in the forest; thus the conversation was kept away from the longhouses that wanted to gain control of their own modernization and industrialization. Over and over, I would have to say, "Forget the Penan, forget Manser, I don't care, pretend they don't exist," in order to talk issues.

No one blames Manser for these difficulties. He is single of purpose and pure of heart. There are tragedies enough in Sarawak, without setting one disaster against the other. Among the natives of various tribes, I found repeated and deep sympathy for the Penan; they are the ones who really know and love the jungle, they are the ones losing it all. Their numbers, many or few, make little difference. To Kayans and Kenyahs up the Baram, as well as to Europe, the Penan are symbols of a kind of purity and a kind of loss. An old Penan from the Magoh River said:

> We know that the dipterocarp seeds are pig food, we do not cut this tree anyhow. The river banks [roots] are what pigs eat, we don't pollute the rivers. Sago fruits, "tevanga" are what pigs eat, we make sure the pigs have their share of these sago products. There is a fruit tree called "tekalet" (acorn). That is pig food, we don't disturb the tree. But in Layun, Apho and Patah those trees are fast disappearing because they are cut down by timber companies. If the companies come here and cut all the trees in the Magoh, there will be nothing for pigs to eat. The pigs will not come here. They will go somewhere else, and we Penan will not have any food. That is what I fear if the companies come here. But as long as the Penan are left alone here, we will have enough food because we care for the forest, we look after it well to provide us our food, our life.

Since that statement appeared in the *Sarawak Museum Journal* of December 1989, the Magoh drainage has been cut. The "little men" have finally been brought into our world, on our terms. Hoorah for systematic collecting.

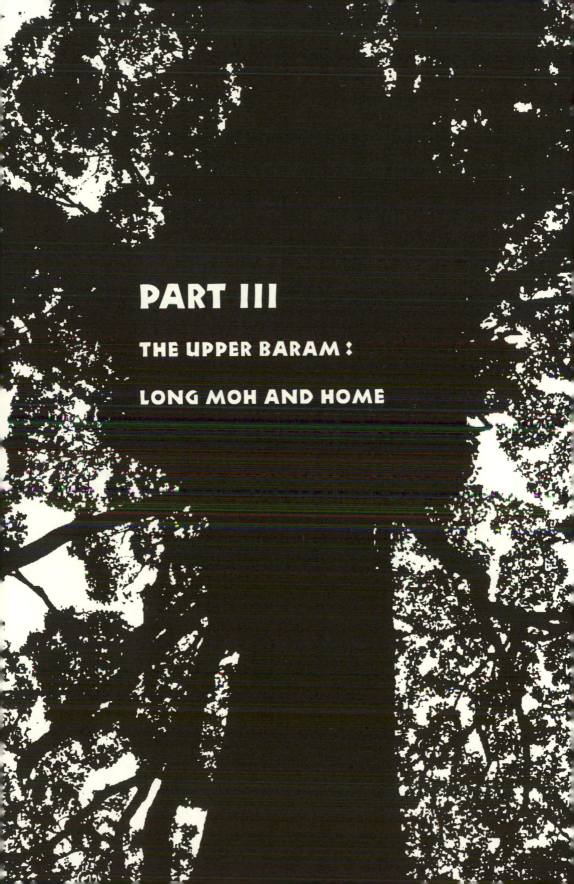

PART III

THE UPPER BARAM:

LONG MOH AND HOME

CHAPTER 10 🦎
LONG ANAP

Sunday, November 18, 1990

Sunday dawned auspiciously at the Tattan Hotel, the two-story building of loose boards at Long San, in the middle Baram. Juliette and I were on our way up the Baram with our young guide Richard, we to the edge of logging and beyond, he to his home. We all awoke to strange shrieks. Forgetting that we could peer through the cracks between boards in the wall, I jumped up and threw open the wooden shutters: the Penan servants were killing and bleeding a pig five feet from the pig sty. The audience of four remaining pigs was excited; they ran back and forth squealing and climbing on each other's backs, jumping against the walls like a crowd in a theater on fire. Then, suddenly, silence — the pigs seemed to know when their brother was gone. A peaceful interlude followed, a time of boiling water, skinning, and cutting. The soul was departed; it was only meat; two piglets curled up in the corner; the curtain was down.

The occasion was the day of little Adrian's christening by the good Dutch Father Jacobse. Adrian, child of Agnes, daughter of the Tattan owners and Philip, a Kelabit from Long Seridan, was a fat and happy three months old. He seemed to have little need of salvation: as the pig was being quartered Adrian lay on his back on the kitchen table, legs waving wildly in the air, while three women laughed and tickled and stroked his penis. Soon he gave a mighty chuckle and peed straight up in the air, then on everyone nearby, the kitchen table, and the floor. General hilarity. Too soon, alas, this innocence passed, and Adrian, sombre and pouting with gas or premonitions, pink and prescient as a pig, was dressed out in hot, stiff clothes and packed off to church.

After church Adrian was a member of a group he had never wished to join, and we too were surprised to be members of a banquet at the

Tattan Hotel. The owners of the hotel, Adrian's grandparents, are the kind of go-getters who alienate people; their house-hotel sits alone on a hill, two hundred yards from the longhouse where they no longer are welcome. They have worked hard and moved up in a modern, or American, or industrial way (by selling cold beer, that is, not by having Penan slaves), at the expense of tribal ties. Their banquet was not a communal affair; it was a gathering of elite outsiders, incuding ourselves, various church fathers and officials, and invited guests, to celebrate the christening of the Tattan grandson. We suddenly found ourselves being introduced to the Penghulu, headman of seven headmen of the middle Baram, and the man who had sold out the middle Baram to Samling. He was from Long San, although he had not been allowed to set foot in the Long San longhouse for two years, since signing the timber agreement. The Tattan owner was also in on the deal; he and the Penghulu were an establishment in exile. They had the power, but could not go down the hill to the longhouse, to home.

Then I turned and found myself face to face with a short, well-groomed man, obviously Japanese. Surprise—Mr. Sei of Tokyo, Samling's regional manager.

Mr. Sei was quiet, polite, punctual in coming and going, abstemious, kind, dressed in a hand-stitched beige safari suit. Syncophants were gathered round, their tongues almost touching his ears. Mr. Sei seemed not to enjoy being rajah of the middle Baram; he held to the Japanese style, discrete, dignified, aloof. Since we had just come from a year in Tokyo, Juliette and I enjoyed the unforeseen pleasure of remembering Japan while sitting in the middle of Borneo. After recollections of Tokyo's Shibuya district on Saturday night, I asked Mr. Sei for whom he worked; he replied that he was employed by Samling; I said yes, but with which Japanese firm was he associated; he said none. He returns to Tokyo twice a year.

Mr. Sei first came to Indonesia in 1964, and has been in Southeast Asia ever since, following the timber. He came to Sarawak in 1976, first to Limbang, then Bintulu, then Long San. In the coming months I had many reports and several encounters with Mr. Sei; he is greatly respected by his employees—down to the yard boys—and is apparently businesslike, fair, and evenhanded. He is married to an Iban woman; yes, she accompanies him to Tokyo. If I were running a timber company, I would jump at the chance to hire this man. He does a job well, and is the kind of person who discourages theories of villainy. The question is, what kind of job does he do so well?

After Mr. Sei had left at the appropriate time, on one beer, and Father Jacobse slightly later on two beers, but still appropriate, the borak began to run freely and we found ourselves, as sometimes happens, with a dwindling number of increasingly inappropriate folks.

So it was that a few hours later around the table there was singing of endless, narrative ballads in Kenyah, which gave way to formal recitations of welcome and praise to the honored Juliette and her soused spouse. The owner of the Tattan, followed by a member of the church, sang long, improvised poems to a ritual choral accompaniment. I was thinking of leaving—if you've ever had borak, you'll know I mean leaving the earth—when the Penghulu of the middle Baram rose up before me, a dark mass in a short-sleeved white shirt, his huge knuckles on the table opposite. My first thought was that he knew we had visited the SAM office in Marudi, and Harrison Ngau, and that we were in big trouble. What if Mr. Sei, the brains of the outfit, had given the word to his goons? I was glad Juliette was beside me; surely between natives and British rajahs, there would be some local gallantry. Standing precariously before us, swaying back and forth on a pendulum of borak, he began to sing. His song seemed different from the songs before, his voice in a new key, his manner directed at me almost aggressively, and yet there was something sweet in his pudgy face, his large, soft tattooed arms, the soft Kenyah language, his eyes rolled up, his head tilted back, as he sang verse after verse.

I leaned to the deacon on my left, and he whispered a running translation in my ear. The Penghulu was composing stanzas about the "knowledge possessed by foreigners" (I thought of Mr. Sei) and of the natives' need for help in their affairs (I thought of cutting the forest). Apparently some of the stanzas were going all the way back to the Rajah Brookes and their interest in native affairs, and that was why he was asking me, now (the truth began to sink in through my rice-grained brain)—me having come so far, and seen so much—to please comment on their situation and give them advice on what they should do. The entire seated company echoed his last verse, and the choral response finished, he sat down. He seemed to have tears in his eyes. Seated, speaking now instead of singing, he repeated his request for advice. I was stunned.

I held his eye for a second, then looked down. In such moments, Asian habits are far superior to American. No one expected an immediate response, spontaneity, glib speech. No one would mind some silence, some space. I turned to the deacon, knowing that among the natives only he understood English, yet all understood tone, and all were watching me. I tried to put my consternation into simple words while keeping face and voice relaxed. "This is very difficult," I said. What is the custom? Did the Penghulu really mean for me to answer, or was this a courtesy to outsiders, a ritual? Was I to respond, actually telling my thoughts, the biggest bore in the Baram? Yes, no, yes. Eyes were on me. I could not keep whispering to the deacon. The Penghulu waited, opposite. I rose, achievement enough at the time, steadied the table which was swaying dangerously, and decided to give exactly what seemed to be asked. With pauses for translation and choral response, I intoned something to the effect that I had found the people of the Baram most kind and generous, and I thanked them, and that their lives were valuable, and their children's lives, and their forests were valuable, and their childrens' forests, and that some thought they were selling too fast, and too cheap, and at a christening we should think of the future and our children's future, and that the whole world was watching and hoping that the natives of Borneo could keep some control over the land of their fathers, and, we hoped, the land of their children.

I sat down, expecting to pay a considerable price for violating every rule of courtesy and taste, preaching the wrong gospel to the very people who had sold out the middle Baram for their own meager profit. As soon as I sat, the Tattan owner began speaking heatedly to the Penghulu; the Penghulu was replying; no one was looking at me. The Tattan owner was angry, others interrupted. There was some loud talk and banging on the table, then the Tattan owner abruptly stood up, said something else in anger, stormed off the patio and disappeared. A loud argument between the remaining five or six people bobbed in his wake.

With guests like me, who needs a party? I leaned to the deacon and asked anxiously what they were saying. Had I made a terrible mistake? Were they very angry?

"They think everything you say is true. They are angry at themselves, and at each other. But it is too late, the middle Baram is finished, and they are drunk and sad."

I wondered if my translator was telling the truth; then an old man, who spoke no English, the father of the owner I believe, came and sat next to me and held my hand, without a word. I was drunk too, and sick, and the party was over; Juliette and I hugged a teary Penghulu and staggered off.

An hour later, we were at a prearranged meeting at the longhouse. As we sat on the floor drinking tea, there was much talk of the middle Baram being concessioned, and of the roads pushing all the way to the highlands. Then my host mentioned the meeting "next week" between Mr. Sei and the upper Baram. "What meeting?" I asked, saying SAM had mentioned an important meeting in mid-December.

"No, it is moved to next Tuesday, November 27th, at Long Moh."

"What's it about?"

"To sign a timber agreement with the upper Baram."

"Are you sure?"

"I have a letter announcing it."

"Could I see it?"

The neatly typed letter was dated November 13, 1990, and was co-signed by the upper Baram Penghulu, Bilong Kuleh of Long Moh, and by Mr. Sei, Regional Manager of Samling Timber. Our friend said the letter was typed in the company office and given to the Penghulu to sign. It called for a meeting with the eight upper Baram headmen on the 27th and listed a four item agenda: to discuss a timber agreement with the upper Baram; to record the identity card number and signature of every headman and head of door (household) in each longhouse; to draw the boundaries of land "between one kampong [settlement] and the next" (clearly implying recognition of native land rights over the entire territory); and other business. My host had received the letter, hand delivered, on November 17. Ten days' notice. In a district with no mail or public phones, and one to three days' travel between long-houses in the middle and upper Baram.

That evening, we were at the longhouse with another native friend who worked with SAM—call him John—who also came upriver with us. I told him about the christening feast with the Penghulu and Mr. Sei just two hundred yards away from the longhouse. "Feast?" laughed John. "It is not a feast if the longhouse does not come!" I spoke of the afternoon's events, the Penghulu's strange request for advice. "The Penghulu probably has regrets," said John, "especially when he's drunk. He knows they signed for nothing. He knows the company has made a fool of him. For that he feels sorry. But he's in their pay now. He went to the upper Baram twice last week in company boats." Then my pickled mind remembered the letter of that afternoon.

"Did you know the upper Baram is meeting with Mr. Sei November 27th?"

"No," said John. "That's not right. The upper Baram meeting is in mid-December."

"I saw the letter," I replied. "It gives the agenda for a meeting in Long Moh the 27th and is signed by the upper Baram Penghulu and Mr. Sei."

John went straight to the radio phone at the mission, and paid for a call to Marudi. Sure enough, SAM had no word of the meeting at Long Moh. But the news must be correct, John figured—that's why the Penghulu had gone up twice the week before, probably carrying bribes. Samling Timber would be pressing very hard for the same agreement as the middle Baram. The SAM office would try to organize a boat expedition to leave Marudi on the 24th; we would go on up to Long Anap, at the edge of logging, as planned, taking Richard home. Then we would join the SAM boat going all the way up to Long Moh, near the top of the upper Baram. We had never even hoped to go that far upriver, days of travel beyond logging, to the most traditional long-houses. The meeting between the upper Baram and Samling Timber would be important, and interesting.

Nevertheless, we felt discouraged as we stood in the longhouse doorway, watching television. Here, halfway up the Baram, was a long-house room packed with twenty-four children and twelve adults watching a Malay romance comedy movie on a television run by a small Toshiba generator which drowned out dialogue. There was wonderful laughter and intensity in the sweltering room, but if the SAM office, with international awards, couldn't even find out in five days about the meeting to sell the upper Baram, were these people ready to take on Japan? Did they have a chance?

Juliette and I walked beneath a waxing moon through a wall of frog and cicada sounds, up the hill to the Tattan. There also, in the family room, everyone was gathered around a generator-powered television. Six people, including the Tattan owner who had stomped out of the banquet six hours earlier, several old fellows, and a nursing mother, were watching a wrestling videocassette: on the screen was Jake the Snake, a peroxide blonde with a Tarzan loincloth and a fifteen-foot python that looked frozen stiff. A very old Kenyah and an old Penan, both with long, stretched earlobes and brass rings, could hardly contain themselves every time the listless snake was lobbed into the ring. They moved forward on the edge of their seats, jabbered and pointed; the owner, in stitches, said they'd seen this video ten times. I could not figure out whether they wanted the snake to eat Jake, or they wanted to eat the snake. That morning, I had asked the Tattan owner if they could still eat wild pig. You have to go far upriver to get it from the Penan, he said, almost $1.50 U.S. per pound. At that price, when a buyer appears, the Penan can't afford to eat it themselves. Now the exiled hotel owner who had signed the timber agreement was laughing at his friends and the snake, but, as in the afternoon, he had tears in his eyes. When the logging comes through, the pigs go away first, then the monkeys, then the snakes.

Tuesday, November 20

> Frère Jacques
> Frère Jacques
> Dormez-vous?
> Dormez-vous?
> Sonnez les matines,
> Sonnez les matines,
> Din din don,
> Din din don.

6 a.m. Why is French running through my head? We awaken to the teeny little melody and electronic chimes of the Fuji clock at the Tattan Hotel playing Frère Jacques. I sit up. Today we go up the Baram, past the logging, to Long Anap.

With Richard and John and a Long San boy, we walk down to the river. They stop and shake their heads. The eight or ten boats of Long San, instead of being beached where they were left last night, are bobbing at the ends of their ropes. The river has risen six feet under clear skies, and as we talk and wrestle an outboard onto a canoe, the water continues to rise, perhaps a foot an hour. John grins: "It rained somewhere." We stow our gear and climb into the thirty-foot hollowed out log barely two feet wide, with two planks nailed and caulked on top of the log. The gunnels are just inches above water; the trip will take about three hours; the first rapids begin a hundred yards above Long San. The river is about as high as we can run it, says John.

There are a few things to know about a river in flood. On a big river, like the Baram, bigger than the Colorado, the problem in flood is not the rapids, which often smooth out into long riffles and standing waves, but the logs floating down, both on the surface and hidden just below, the ripsaw eddy lines, the sudden brown boils and surges, and the general bank-to-bank speed. If you tip over, you don't swim to a beach or a gravel bar but to a four-foot high vertical bank of twisted roots and ferns which you're floating past at five knots. You could grab handfuls of orchids for miles before getting out. The good news is there are very few crocodiles above the rapids.

Our hollowed log, roaring full throttle out of the Long San eddy, is suddenly lifted two feet and shoved twenty yards to the side by a boil; it yaws and skids back into line for the rapids. This being upriver Borneo, of course, no one has ever heard of life jackets, much less dry bags for cameras, and our sole defenses are the Suzuki 30 outboard motor and, as water breaks in our faces and the boat slides across currents, continuous laughter and shouts of "No problem." At the helm Richard is relaxed, eyes straight ahead, his hand resting light on the throttle—a cowboy's hand on the pommel of a runaway bronc.

They know the river well, where to take each rapid, where to hug the bank, and the young boy in the bow—another cousin on our payroll, apparently—points out trees rolling down on us in the brown boils. We bounce over the odd log now and then, and the motor falters in debris, once or twice in the middle of four-foot waves, but as we begin to drift and roll backwards down the rapids Richard gets it started again and we surge forward. We roar up the eddy beside a long, fast rapid, then try to cut into the slick at the top, above the waves. We slow almost to a stop, engine wide open trying to push our thousand-pound log forward, the river racing by, trying to push us back. For a second we are still, vibrating, water and logs rushing past. One log too many at that moment, a twenty-mile-an-hour collision, and you're swimming. Then the propeller begins to grab—we can feel it, half a mile an hour forward, two, three—and we are above the tongue, racing across the flat water to the next set of waves. Juliette, who has canoed big water from Idaho's Salmon and Selway to Glacier Bay, Alaska, is white from nose to knuckles.

Late in the morning we glide into an eddy with a huge notched log sloping down to the river; buildings are visible at the top of the steep bank. Richard's eight-year-old sister Twyla stands on the log; the baby gibbon on her shoulder hides behind her head at the sight of the boat, holding her hair across his face like a veil, peeking through.

We are at Richard's house, Long Anap, above the logging. He has not been home for over two years. We climb out of the canoe, weak kneed and faint hearted, and carefully ascend the slippery log. He and little sister Twyla speak softly. People appear from the trees; we cannot tell who is Richard's family and who is not, the greetings are so quiet and shy. John says Richard's mother is in the longhouse, and someone is fetching his father, who is hunting nearby. We walk through the leaves to the clearing.

Long Anap is a small longhouse near the top of the middle Baram, living as it has for a long time with the addition only of outboard motors, zinc roofing, a generator, and a school. Around them is cultivated land and virgin forest. This month, however, a few miles behind the longhouse, Samling's road is creeping along the ridgetop, two hours walk away. Within two months, logging will begin. Long Anap is in Block 3 of Samling's 1991 concession. The people of Long Anap protested the logging and blockaded Samling's road in September, and are about to try again.

Soon we are seated in the deep shade of the porch, drinking weak, sweet coffee in a light breeze, fifteen feet above the ground. Beside us on the floor, five women are weaving mats; five children run about, two are nursed, and a teenage girl is helping. Behind them, two adjacent doors lead into rooms of the longhouse; one door has a Harrison Ngau election poster left over from the October campaign, the other is handpainted with a Marlboro cigarette logo and a three-foot-high reproduction of the label for Lone Star beer.

The women are weaving a five- by eight-foot cane mat, and invite Juliette to join them. She sits between the women; they share no language, but are all smiles as they work side by side holding the mat in their laps, hands flashing back and forth like a school of fish. One big woman behind them lays twenty-foot lengths of a canelike vine down on the porch, and using her fingernail, slits the cane into quarter-inch strips. The mat will be her family's. The cane strips, and the woven mat, are blonde, shiny and stiff. The women distinguish about fifteen types of vines, and call this one *tepo*. We are told it is specific to this kind of mat, for the high use area in front of the family's door. They say it will last five years. The border is an intricate counterweave. The vine gathering took a day, yielding much more than this mat will use. The rest is stored in lengths on a rack beneath the longhouse. The mat was begun yesterday and is almost finished. The porch weaving, of course, also serves as social hour and day care center while other mothers work in the field. They are speaking to Richard, and he translates. *Tepo,* one of three or four common mat materials (each prized for different qualities), is getting harder to find, they say. After the logging, he tells them, it will be gone.

Modernization theories say that soon they will be able to go to Marudi and buy mats with cash. Or a store will open in Long Anap, mats trucked up on the road. Either way, some of the $200 of gas used for the trip will be profit for Shell Oil at Miri, which is not native owned. Subsistence economies, as far as I can tell, benefit only the people living in that locale. Is that why my country calls them underdeveloped?

A "softwood" fire with imperceptible smell and no sting to the eyes smolders in a coffee can to keep bugs away; one woman chops betel nuts; they hand us raw sugar cane to suck with our coffee.

The teenage girl has injured her foot at the river; it is deeply scraped over a wide area, and bleeding profusely. Her mother asks for medicine. We warn the girl that we have only stinging iodine (which doubles as water purifier), but no expression crosses her face as I pour iodine on the open wound. She seems a bit shocked, then laughs and runs away. One woman and one boy also have swollen and wrapped toes, and I think of Redmond O'Hanlon up the Rajang among the Iban, saying that foot injuries and their infections are the most common form of early death. An open wound on a bare foot in the jungle, and in the chicken-dog-dirt ecology beneath the longhouse, is trouble. Antibiotics are rare and avidly sought.

The benefits of modern medicine, however, must be measured against the problems of population growth, which may already be straining the carrying capacity of the jungle. In every longhouse up and down river, I asked all who might know for estimates of population growth in the Baram; most old men guessed that numbers have at least doubled or tripled since World War II, partly from medical treatment of individuals, largely from a drop in infant mortality and the near eradication of malaria and cholera. In some cholera epidemics of the early twentieth century, 80 percent of the natives died (similar to smallpox disasters among Native Americans). Since major migrations of tribes have occurred in the last two hundred years, and longhouses are always splitting up or moving, no one can possibly assert what might be a baseline or optimum or stable population in the middle Baram living traditionally in the forest. Almost every elder and headman I asked could name three or four longhouses which have tripled in population within his memory. Two decades ago Long Anap had grown to about fifty doors (households); the shortage of land (and the usual political disputes) forced many to move over to the Tinjar. Now there are twenty-seven doors at Anap, or about two hundred people actually living there, although the tribal roll includes about eight hundred. Many are usually away, at school, working, visiting; only at Christmas time is a longhouse full.

Even without logging, therefore, it seems likely that forest resources in the middle Baram might be scarce, *tepo* increasingly hard to find, and that a hunting-gathering culture in the Baram would soon have faced severe challenges to its traditional lifestyle. That point had not quite been reached, however, when the logging came. Harrison Ngau believes that the Baram population has dropped in the last few years, with family planning and people moving to town. We do not know what stable population the land could have supported.

A longhouse is never finished, and this one is no exception. As in the forest, growth and decay are simultaneous: always someone is adding a room, something else is falling apart. The family space in most longhouses is walled off from the porch with the door normally shut. Handsawn planks form the walls, now competing with plywood and fiberboards brought up by boat. The floors, empty of furniture, are big and dark, foot-polished hardwood with a square or two of linoleum (easy to clean) added to the sitting-eating areas.

The outboard motor has brought one huge change upriver: zinc roofing. The old roofing was made of hardwood shakes (or leaf thatch), and was dry and cool, but required enormous time to construct and, surprisingly, unless it was made of ironwood (*bilion*), did not last that many years. The zinc sheets save hundreds of man-days of labor, though they radiate heat and therefore individual rooms must have flat mat ceilings under the huge, high roof. This cuts down ventilation. The old, open ceilings allowed more flow of air the length of the longhouse, probably the reason for the traditional raised ends of roofs in Indonesia and Malaysia, crcating updrafts.

A four-by-eight-foot zinc sheet is $3 at Long Lama; and a round trip costs $200 in gas (the number of sheets per trip varies by boat size, but the loads would not pass any safety inspection I know of—canoes crash up through rapids with inches of freeboard). A typical longhouse door (unit) is roofed by sixty sheets of zinc. If the load comes up on one boat, the total cost of a roof is about three months of a wealthy family income (maximum $150 a month) working five acres of rubber. The rubber income, and the zinc-gas expenditure, is a major cash transaction in most families' lives. Naturally, one member of the family working in a logging camp making a minimum of $150 a month adds greatly to their buying power.

The architecture of the longhouse is beautifully adapted to life in the tropics. Six to fifteen feet up on stilts, the house is clean, dry, and ventilated, and separate from domestic animals, mud, parasites, and flooding rivers. Dogs may or may not be allowed on the porch; many porches have gates at the top of the ladder to keep them out. Some longhouses have rooms beneath the porch with dirt or cement floors, some have storage areas only, some are just open rows of pillars, stored boats, and sleeping dogs. Long Anap has walled storage spaces on the ground leading to a stairway up into each family room. There is no public, outside entrance to the porch—a bit unusual. The porch is common to all doors and open; the doors are private, leading into family rooms sometimes subdivided into sleeping compartments. Nonfamily persons knock on a door and await permission to enter. With almost no furniture (all sit on the floor), the rooms are large, light, and airy.

At the back of the longhouse each apartment has a little walkway to its kitchen, an addition—also on stilts—sticking out to the rear. The walkways and kitchen sheds are roofed and joined to the common longhouse roof; the open-sided walkway to the kitchen gives a pleasant view of the other buildings and may be joined to the next door walkway if desired, to borrow cups of sugar (usually next door is family).

Cooking smoke, fire dangers, odors, and insect-attracting foodstuffs are thus separate from the living quarters. With no walls, slop can be pitched off the kitchen porch. Yet no trash (besides plastic) accumulates; pigs, chickens, and dogs recycle the garbage, and some waste is composted in the fenced gardens.

The kitchen area is also a bathing space (supplementing the river); one stands on open hardwood planks, soap and towel on a shelf, and pours ladles of cool water over the head, usually wearing a sarong (public nudity is taboo almost everywhere in Asia, and most kitchens are partly visible to neighboring kitchens, ten feet away). The runoff washes the kitchen floor and drains between the planks to the ground.

Every observer has remarked on the personal cleanliness of the Borneo natives. The morning and evening baths in the river spare no suds, especially on the hair, and the only person I ever smelled was myself. Eating and sleeping in longhouses we never encountered vermin, although children playing in the dirt sometimes pick up parasites and have their heads deloused or shaved. Floors are constantly swept and frequently washed. Even mosquitoes are mild, nothing compared to Montana in July. Leeches are absent near the longhouse. In all, it is a controlled, clean environment, with considerable privacy (at the family, not the individual level), or communality, as you wish. Upriver natives have five times as much floor space as the average Japanese. Many of the longhouses have beautiful plantings—bougainvilleas, hibiscus, and half a dozen other flowering shrubs—in addition to the pleasant and useful banana, rubber, palm, and fruit trees scattered about the longhouse (mixed; never as plantations), so that the view from the porch or the semi-open kitchen can be lovely.

We settle down on the floor of Richard's home. The room is empty and spacious; later I measure it thirty-five by twenty-four feet, the walls ten feet high to the open rafters. Above the rafters there is no ceiling, no wall; the room opens to the zinc roof twenty feet up. The effect is that of a cathedral ceiling with open air gables; since the open gables extend the length of the longhouse, body heat can rise and flow out the ends of the building. Hanging on the wall opposite me are two large hats for working in rice paddies, three woven baskets, twelve plastic cups, a saw, four strainers, one grater, two dozen clothes hangers, and a framed print of Joseph and Mary bending over the manger in a field with gothic ruins in the background. On the wall behind us hang a satchel of fake leather, a small ornamental canoe paddle, and a clock with the face of Jesus, the hands coming out of his nose. Richard's father appears, in from gathering wild nuts; his mother had already returned from weeding their hill paddy of rice. Children are coming from all over.

Richard's father is forty-four. His mother is forty. If I told you how handsome his father is, or how beautiful his mother, you would not believe me. Their grace and modesty calm the house, and in their presence the children, as in most upriver homes, are quiet and considerate, serving tea or rice as needed, otherwise grouped in dark corners staring, breaking into wreaths of shy white smiles when challenged. We reach into the pack and pull out our video camera. Do you want to see yourself on television? The magic words, translated, bring a few forward. I turn the camera first on myself, then on the oldest and boldest. They scatter, but creep forward for the playback. One looks through the eyepiece. Exclamations. Soon they are delighted to be filmed and to crowd around for viewing. A two year old will not loose her grip on the eyepiece, glued to the tiny tube. Never once, with our precious camera set on the floor of the longhouse, does someone grab or push. The hilarity spreads to Richard's grandmother. I have filmed her; as she bends to the camera on the floor her earlobes and brass earrings drag on the linoleum. As she leans to the eyepiece, I push "play"; she cackles and the children howl. "I am not so old," she says, straightening up with a toothless grin. "Yet that is me." The two year old crawls back like a leech and fastens onto the eyepiece.

Richard's father, Kenyah from Long Anap, is related to the assistant headman and cousin to Thomas Jalong of SAM. His mother, a Penan, is sister to the headman at Long Bangan. She converted to Christianity for her marriage. Her name is Flo, short for Floral, a translation of a tribal name. Their eldest son, John, twenty-three, is a clerk in Miri, studying to be an accountant. Our friend and guide Richard, twenty-one, has no idea what he wants to do after he has made his millions in timber, but would prefer to be upriver unless it's ruined. Another son who followed Richard died at the age of four. Then came three daughters: Amy, a pretty and graceful eighteen year old engaged to a boy from Long Selatong; Judith, sixteen, a student at Long Lama; and little Twyla, eight, in a Teenage Mutant Ninja Turtle T-shirt, her pet monkey clutching her back. Twyla came as something of a surprise, we gather, and at first she had a tribal name. Then about two years ago Flo saw the movie *White Nights* on video, liked the name of choreographer Twyla Tharp, thought Twyla was a kind of twig and a pretty sister name to Floral, and renamed her daughter.

The men and guests settle in a circle on the floor for lunch. Richard's mother brings a white enamel pitcher of water, and sets it in a tin basin on the floor. One by one the men get up, walk over and, holding the pitcher in the left hand, pour water over the right hand above the basin, rub the hands together and dry them with a rag on the floor. When we are all washed and seated, the girls set banana leaf balls on the floor in front of each of us. We unwrap our banana leaves and with our fingers eat the sticky rice off the leaves, dipping rice balls in the soup of Spam chunks in a dish in front of each of us. Flo and elder daughter Amy serve, and also eat a bit with us; sometimes a child finds a mouthful.

After a rest, we tour Long Anap with John from Long San, Richard and his father, and a growing number of adults, children, dogs and roosters, and one pig — finally an entourage of thirty or so.

Long Anap, not particularly prosperous or progressive, without significant government grants or mission funds, offers a fair view of life two months before logging. We climb down the stairs to ground level, and walk out between the stilts to stand between two buildings. We are in between Anap's three small longhouses — with six doors, four doors, and eight doors respectively — all in a row facing the river, set back from the top of the bluff. As is usual, big trees are left standing near the longhouse, so we are in the shade of rainforest hardwoods, three tapped rubber trees, and several twenty-foot-high coffee plants with red berries. Four or five other scattered buildings, separated by fenced gardens, also have families. Uneven ground, political as well as topographical, has come between families, forcing separate longhouse structures. Between the longhouses are two large, fenced gardens, and all about are numerous outbuildings, perhaps ten or twelve granaries on stilts with flat stones set on top of the stilts to keep mice out. There are some outhouses and shade sheds; at one, we chat with an old man caulking a new boat. We peer in through the windows of the locked chapel near the river, shoot a few baskets with three boys on the cement basketball court, and walk out the end of the settlement to the white frame schoolhouse and its open playing field.

The children go to primary school here and to Long San for junior high through form 3, both compulsory. Forms 4 and 5 at Long Lama, and 6 at Miri or Marudi, are optional. Instruction, which used to be in the Queen's English, since 1983 is in Bahasa. The Malaysian government is quite serious about public health care and education.

The grounds in between and about the many buildings are filled with gardens and planted trees and shrubs, most of them fruit bearing and owned by families. Here, much more than in primary forest two hundred yards away, one feels the presence of "jungle." The growth is everywhere, green, ripe, prodigeous. Monster fruits, durians, red rambutans, tiny bananas hang from dozens of types of trees, and the various gardens are studded with greens. On the slopes behind the longhouse, hillside rice is greening, to be harvested in January; some corn, cabbages, and cucumbers are mixed in with the rice. We are surprised to learn that their diet has hardly changed since the outboard motor — "90 percent the same," says John — because food from Marudi is so expensive. The Spam we had at lunch was a luxury because it was an import, but they find sufficient meat in a primary forest. We are looking at a subsistence economy: local farming, hunting, and gathering. It doesn't look so bad to us. The terms "underdeveloped" and "third world" do not at all suggest the quality of this life.

Like most longhouses in the middle and upper Baram, Long Anap is on a flat between the river and a hill. The land traditionally belonging to the longhouse extends from the river back to the top of the ridge behind the settlement, and also across the river in a similar swath up to the crest of the opposing ridge. The boundries to each side, up and down river, are halfway to the next longhouse. This means, of course, that all the forest, at least up to the first ridge back from the river, belongs to one longhouse or another, and that the traditional boundary is the one recognized by Samling's letter of November 13. Behind that, deep inside as they say, is Penan country. Within this rectangle of tribal property rising on both sides of a river, each longhouse recognizes three kinds of land.

First is cultivated or cleared land, worked by a family and belonging to that family as long as it is worked. Right to the land is established by the act of clearing, and ownership extends through periods when the land is left fallow during the normal rotation of cultivated plots (shifting cultivation). Here is where each family grows its rice, corn, cabbages, cucumbers, sometimes peppers and planted trees. At Long Anap, about 30 percent of the tribal land is under cultivation.

The second category is tribal forest, again about a third of the area. This is the forest open to common use, including clearing plots for new cultivation, felling trees for boats and house planks, collecting vines, and any form of gathering or hunting.

The third category is a forest preserve, a protected ecosystem. Hunting and renewable resource gathering are allowed, but cutting is prohibited unless the longhouse votes to cut, for instance, a certain tree for a new longhouse beam. Regulations for gathering in the preserve are very specific: illipi nuts are to be left on the ground for the pigs. At Long Anap, the forest preserve is also about a third of their land area.

On paper, the forest preserves can be protected from logging by each tribe. The natives may go through a process of application and designation of a forest preserve within a certain period after a timber license is granted, that is if they hear about the license to cut their ancestral land (without mail service, there are few subscriptions to Kuching's *Sarawak Gazette*). To date, not a single forest reserve request has been granted. Long Selatong, for instance, in the middle Baram, wrote many letters within the designated period, and was told "your application is being processed." Then the area in question was logged by Samling. After the logging, the government informed Long Selatong that a forest reserve could be established only in primary forest; therefore their application was being rejected. People from Long Selatong blockaded a road at that time, on their ancestral land, only to be arrested under the new forest law for obstructing a timber company.

Across the river, Lambir told each longhouse in its concession that they could protect a hundred trees by applying to the government. Long San applied; the government never answered the letters or subsequent inquiries. The trees are gone.

Damage to cultivated land by logging, usually by roads, is compensated; that is a main provision of each timber agreement. Whether the amount is reasonable we will hear this evening and the rest of the week. Basically, once a timber concession is granted in Kuching, with absolutely no consultation or even direct notification of the natives, no part of their forest is protected. None. So far, no one has bulldozed a longhouse itself, but many precious nut trees and even planted fruit trees near the settlement, as well as in all sections of the forest, have been taken—no matter what the provisions of the law. This news of actual practice has to travel from longhouse to longhouse by word of mouth, since it is certainly not the picture the company and government paint of what is coming, nor can it be imagined from reading the timber agreements.

In the light of these facts, facing productive gardens that would seem to satisfy so many needs, and well aware these are not nomadic Penan, we pause at the edge of the clearing, look back over the settlement, and ask Richard to translate to the dozen or so adults and children and dogs and pigs surrounding us. "What do you get from the forest? What will you lose when the forest is cut?" Richard asks, and adds that the forest may be cut within the year, though they cannot imagine that, even when told. Men and women talk excitedly. Richard's father, tan, ruggedly handsome in his blue shirt and baggy shorts, looks slowly about at the hillsides, and speaks.

First, he says, they will lose the timber. Timber for boats and housing, posts and planks. They need only trees, a chainsaw, gas, and nails. We have watched several boats being made: two men select a tree, cut it, then slice the trunk in half lengthwise with a chainsaw, taking half for a canoe and half for planks. With hammer and chisel, a man hollows out the half log to a canoe thirty to fifty feet long, two to three feet wide, one to two feet deep. The hollowed-out shell has walls about two inches thick. This is all done by chainsaw, adze, chisel, hammer, and hand planes. The tools have store-bought metal blades, ancient and well honed, set in beautifully shaped, homemade hardwood handles. To make planks, a man on top of the remaining half log walks its length, holding a chainsaw straight down, and cuts a straight plank one and a half inches thick. They build up the sides of the boat with these planks, about ten inches high, set lengthwise on top of the hollowed log. The planks are nailed and caulked and at bow and stern, bent together and fastened to posts fore, aft, and midship. The wood is meranti, the caulk is damar or jelutong from those trees, and the boat can be finished in a week, although it more likely provides afternoon puttering for months. In every longhouse we visited, two or three boats were under construction in deep shade near the river, where on sleepy afternoons a man could chip away at his canoe.

These are river and fishing people who need boats; outboard motors are shared. As we had learned, such a boat can run hundreds of miles of rapids for years, and is tough enough to be driven right up on a stone beach. It weighs about a thousand pounds, has a square stern which can take up to two forty horsepower outboards, and the total cash cost per boat is $4 for chainsaw gas and nails.

All they need is a meranti tree two and a half feet in diameter. Unfortunately, meranti two feet across or more are just the trees the timber company will take at first cutting; and with absolutely no part of the forest protected, they will take them all. Housing planks are also meranti, and are also cut by chainsaw and hand planed on the premises. Housing costs are limited to gas, nails, and zinc roofing. The posts are *bilion,* a wood so hard even logging companies, so far, leave it alone.

What will they do, Richard's father asks us, without trees for boats and houses? We have no answer.

As well as timber, he says, they need rattan and especially three or four other woody vines and canes. These are used for mats, woven baskets, hats, chairs, and furniture. Although the floor is the usual sitting place, many doors have at least one chair for the aged, and cane furniture can be sold in Marudi. Other vines are also used for every manner of lashing and rope. I have seen a two-thousand-pound boat held in current by an old rattan line. Richard's father is speaking to them about different vines and their uses; several are gesturing as they try to speak directly to me, tying knots or weaving baskets in the air. They easily name six vines for different specialties and seem to know dozens more. John says the vines will soon disappear. Several men are surprised; why does the company want vines? The company doesn't, John explains, but the vines catch and bind and get in the way, and are torn out by the falling trees, dozers, cables, and logs being skidded to the road. Different kinds of vines invade secondary forest. Two years after an agreement has been signed, and two months before logging begins, this is clearly news to some of Long Anap, and they look depressed. Richard's father smiles and speaks softly but clearly. "I told you so" is John's translation.

Third, says Richard's father, the animals. Most of Long Anap's red meat comes from the forest. Pigs are found only in primary forest, where they forage certain fruits and illipi nuts (hence illipi gathering is prohibited in the reserved area, to keep the ridgetop on the pig migration route). Since pigs wander, following fruit blooms, it is hard to say they have increased or decreased over a few years in a given district. Long Anap still found ten or twelve wild boars last year. By now, the pigs have certainly fled the logging sounds. Most birds and monkeys will leave a cut area, they say, although some deer may remain. An older man speaks; he has relatives far downstream. Longhouses there find nothing like their old supply of meat when the primary forest is gone. He takes off the straw Kenyah hat with reed bill and two hornbill feathers, wipes his head and spits a stream of red betel juice.

Fish, says Richard, supply some or most of their meat, depending on the season. The Baram here alternates between clear and brown during the rains. Downstream people say that after two years of solid brown (from logging road erosion), the fish population declines. Richard tells them about the Tutoh; in two years fish completely disappeared from the Long Bangan tributary, and now it is so muddy they strain the water through T-shirts before boiling it for drinking. The government has said they will do something about the water supply, build a tank, but nothing happens. The logging is past, and the erosion only gets worse. John says when he was growing up at Long San, one cast of the net often caught dinner for a family of four; now four to six hours of fishing produces the same amount of fish. The Long Anap people listen closely and look at each other; the two roads are right now crossing their boundaries, along the ridges on either side of the river. They report that only last week the road on the opposite side crossed a new tributary, and the people in Long Palai say the stream went muddy for the first time ever (some streams are clear even in rain), and there have been no fish caught since. This news, too, appears to take a few by surprise; they knew fishing was affected, but not so quickly, so severely.

Richard's mother has joined us and speaks of collecting fruits and materials in the forest: durian, jackfruit, rambutan, wild sago, damar, jelutong, nyatoh. Not all can be cultivated. Others mention jungle products which can be sold for cash: illipi nut, for example, the pigfruit tree, is protected by law but loggers sometimes rename it meranti and saw away. The illipi has oil for preservatives, a kind of chocolate, and the nuts, if not left for the pigs, can be marketed in Marudi: one tree in one season can yield almost $400. Such fruit and nut trees can become "owned" by families and are prized possessions. There are at least a half dozen money-earning types of trees (including rubber) wild in the district; even if not cut, John reminds them, some may be among the 30 percent of trees damaged during logging operations. One middle-aged man who had looked surprised during these conversations spits and leaves the group.

I am beginning to sense, as the logging approaches (right now, three miles away), how little these people know, how little they are prepared. But am I being too extreme? Then I remember a visitor at Long San, a European businessman many years in Sarawak, who had watched the whole process upriver and down. "What happens after the logging comes through?" I had asked. "The end," he replied. "Finished. Nothing." What about plantations? "Up here? Too far from market. Rapids, poor soil. Besides, plantations are real coolie work. Now at least these people are independent. They work for themselves, and they can move up in life. After the logging, with wages and plantations, they are stuck. They'll never move up. It will be the end. Don't quote me or I'll be thrown out of the country."

I stand in the midst of people who have much less apprehension than the businessman, and much less knowledge. Beautiful, kind, hardworking, clean people. I remember Rajah Charles's words to the native council in 1915. "You people of the soil will be thrown aside and become nothing but coolies and outcasts." Many natives think they are renting land out for a while, says John, and that a road will come to their longhouse, and a few trees will be taken, and maybe afterwards things will be the same but they'll have more money and a road. It is precisely what most Indian tribes in America thought when they were paid for land taken: a rental fee, how nice. Soon the whites would go back to wherever they had come from, and things would be the same.

Richard has a faraway look, staring at the ridgetop downstream. He has been through it all on the Tutoh. Like many modern boys, he knows more about some things than his father; now it is all coming home. Last week when the wind was right, says Richard, and the roosters were quiet, from the porch they heard machines.

Richard, John, and I make quick calculations. If they lose 80 percent of their meat supply, all their free building materials and boats, their wild fruits and nuts to supplement garden vegetables, and perhaps half their cash income, their economy is not affected, it is destroyed. And after the logging has passed and their sons no longer bring home company paychecks . . .

Most ominously, however, they are changing to a new kind of

living, from a free economy to a cash economy, in which they are as lost as I am in the jungle. The compensation will be in cash, yet they have no basis, no experience, on which to judge cash value. What does a house cost, or a boat, built of imported materials or by someone else? Before the logging, thirty dollars sounds to these people like a substantial sum; it is, when only five percent of their economy involves cash; it may be a month of cash flow. Cash to most is really pocket money, except for the occasional outlay for roofing and the pooled resources for a motor. But if they have to buy all their baskets (plastic baskets?) in Marudi, thirty dollars will seem like nothing. It is exactly at this point that the injustice so easily occurs: a Japanese trading firm worth billions sits down, through its representatives, on the longhouse porch with Borneo natives, and offers cash for their old world to bring them into a new world they cannot imagine. The medium of forest resources, which the natives know so well, is not the language spoken. That language is dead. The terms are cash; in that medium, the natives are illiterate. The Japanese are not.

The question that has nagged me for months is just how they got into this mess; it was not quite at gunpoint. Now, finally, we have enough guides and friends and intermediaries to ask the right questions, and tonight, at a meeting in Richard's home in the Long Anap longhouse, we will hear just how and why the middle Baram agreement was signed.

AFTER DINNER with Richard's family, the gas lantern is lit and hung from a beam. Under its light, the circle of people on the floor grows larger by the minute, and soon almost thirty people are seated in the room, spreading to the dark edges. Like every longhouse after logging has come, Anap is split between those who stood by an acquiescent headman and those who formed an open opposition. In the room are the leaders of the Anap opposition, as well as several leaders from other longhouses in the district. Men are in the inner circle; quite a few women are back against the wall. I had asked at the Marudi office if any headmen had told SAM directly how agreements come to be signed, if any participants in the alleged payoffs had finked on the company. They said none had, reminding me that monthly payoffs can be suspended, but they added that there are no secrets in a longhouse, and the pattern was by now well known. As it turns out, at least one eyewitness participant in the middle Baram will tell the story.

The middle Baram has seven longhouses at present (from time to time, longhouses subdivide or are abandoned). From downstream: Long San, the largest settlement, with a mission and a government school; then, only an hour above Long San and Lambir by boat, Long Selatong and across the river, Selatong Ulu, two quite prosperous settlements; then, several more hours up, Long Apu (and Julan, almost abandoned), Long Pelatai and Long Anap, smaller and scattered; and finally, Long Palai, a thousand people and several big longhouses, almost at the big bend of the Baram.

The Penghulu from Long San, who lives now just downstream in Long Akah, where he is part owner of a store (Long Akah has a small dirt airstrip and the farthest upstream Chinese stores in the Baram)—he of the tearful invocation at the banquet on Sunday—is Penghulu of the entire district. He is legally recognized by the government and receives a monthly government stipend; the same is true for the head-man of each longhouse. All are "elected," from the aristocratic class, although the procedures for nomination, election, and removal, not to mention enfranchisement, vary from longhouse to longhouse. Since the 1970s headmen have been ratified by the government's district offi-cer. Before then, headmen simply served or ruled their longhouses. Their recent elevation to legal status, however, plus the monthly sti-pend from Kuching, gives the government a new relationship with each longhouse. Headmen have a vested interest in working with gov-ernment. They can be offered more, or less, and they can sign things.

In December 1986, says a thin-lipped, bright-eyed Kenyah man who will speak with evident wit and irony for an hour, Samling called a meeting at Long Na'ah downstream, for all the headmen of the middle Baram. Some others went along as unwanted audience. The stated pur-pose of the meeting was to discuss a road to Long San—where it would go, compensation for damage to crops and fruit trees, and so forth. A road is eagerly sought by most longhouses, for it promises easier, more dependable, and faster access to the outside world. At the meeting no mention was being made of logging, so this man and several others interrupted from the audience. Why doesn't the government build the road to Long San, they asked. The government would take too long,

Samling replied; we'll do it for you. Why? Maybe we'll rent it out, or if prices ever improve, maybe we'll take some timber. The company offered $5.75 per square chain of cutivated land crossed by the road; the meeting asked for $11.50. The headmen wanted to sign at this point, but the audience prevented it.

After a few weeks, a company agent picked up each headman, alone, for another meeting, this one unannounced. For a headman to go alone to a meeting concerning longhouse affairs, without a single adviser, is unusual; in the old days, of course, his office was not a government post but a coalition of longhouse interests. At Long Na'ah again, but this time in private, the seven headmen were bribed to sign seven identical agreements with Samling for roads, logging, and compensation. When they brought the agreements home, the longhouses rejected them. People knew the compensation to the longhouse could not make up for the loss of their forest. This happened again in 1987, and again in 1988. Each agreement was the same; each time the headmen and Penghulu signed, and the longhouses refused to ratify. The current logging is proceeding under the agreement signed January 15, 1988, by the Penghulu Lanyau, and the various headmen, including the headman of Long Anap. He is not in this room. Since the headman has the legal right to sign documents, and since "there are no native land rights," the timber company does not have to have longhouse ratification. It would just make things nicer.

Under the kerosene lamp, beside my candle on the floor in front of me, is a copy of the Long San agreement which I had managed to acquire (copies are *not* made available to the longhouses or the public). It is signed and thumbprinted by the Penghulu, the Long San headman, and the two Long San committee members. The agreement has nine pages; about one-third of the middle Baram headmen can read English, and whether any can read legal English I could not ascertain. "The company did a lot of explaining," said one eyewitness. "Couldn't the government provide a lawyer to help them?" I ask of the circle under the kerosene lamp. My question, translated to the gathering, brings a round of hearty laughter.

I say that I understand bribery, but please tell me more about why the headmen sign. At the closed meeting downstream, says the thin-lipped man, the company told the headmen that longhouses have no land rights, that the timber concession had already been granted to

Samling, that nothing could stop either the roads or the logging, and if they wanted any compensation at all for themselves, above or below the table, and for their people, and a feeder road to the village, they should sign now. Last chance. Later, when I spoke to Samling officers at their headquarters in Miri, they repeatedly stressed that they had not paid "compensation" but rather had made "presents," freely given and freely accepted. Their generosity was voluntary. "Why," I asked, "does Samling bother to give 'presents' at all?" "It is hard for some of them," I was told by an official. "The fishing is hurt."

The 1988 agreement in my possession speaks of the feeder road as "consideration" for entering tribal land, of payment for cultivated land as "compensation," and of a "goodwill" payment to the entire longhouse.

I ask the thin-lipped man if anyone knows for certain exactly what the headmen were paid. Yes, he says, and introduces another man who has come to this meeting to tell me himself. Trust me—he knows. Each headman, he says, was personally given over $200 cash (500 Malaysian ringgit) at the meeting, and a promise of up to $200 a month for the duration of the logging (about three years), if he would sign. These bribes, of course, did not appear in the agreement, and were not reported to the various longhouses.

In the agreement with Long San (identical to the other middle Baram agreements), the longhouse itself gets five considerations: (1) a feeder road to the settlement (which, of course, they or a government would have to maintain after the concession is finished—a mud road through steep rainforest with little traffic); (2) a promise that the company "shall use its endeavour to employ people of the said Kampung [longhouse] *PROVIDED THAT* they have the qualifications" (emphasis theirs); (3) a "goodwill" annual payment to the longhouse (for Long Anap, about $1,000 a year); (4) payment to the longhouse of $0.20 per hoppus ton of timber removed from tribal land (as of December 1990, this had not been paid in the middle Baram, and downstream people believe this sum is never calculated honestly; (5) payment to individual owners of $11.50 per square chain of cultivated land ruined; and varying, specified payments for each ruined fruit tree: durian, cocoa, coffee, rubber, and others.

Attached to the Long San text are two curious documents: a letter, and a list of seventy-three signatures. First is the letter. "Sir: We the inhabitants of Kampong Long San do hereby agree that our representatives whose names are . . ." Since the "that" clause never reaches a verb, the text wanders into a jungle of syntax. The thrust is apparently to confirm the identity of four representatives and four witnesses. "We . . . agree," however, without any meaningful qualification and followed by three separate pages of seventy-three signatures, with no heading at all, seems to imply consent to the agreement by all the longhouse signees. That is not the case, I was told repeatedly and angrily; the signatures were obtained for another reason and attached here (I remember an item in the agenda for Long Moh: identity numbers and signatures of all heads of household . . .); in addition, several signatures are palpable forgeries (the alleged signees wrote their names for me, in a totally different style, and showed me other documents they had signed).

Only a few people at Long Anap have ever seen their agreement; they did not want it signed, thrice refused to ratify it (although Samling's three attempts imply that longhouse ratification is part of making the process legitimate), and now they can only accept whatever funds are disbursed. Their forest, belonging to the longhouse since they came to this spot perhaps eighty years ago, and guaranteed to them under the system of *adat* which may be thousands of years old, will be gone.

The people crammed into the room have grown quiet and serious during these recitations. There is a lull. I explain to Richard and John that I would like to ask a series of questions, just for the record, and record Long Anap's response. They relay the message, and people lean forward. "What benefits do they expect from the logging?" The room buzzes; they gesture to each other. Quiet. None, Richard says. I mention the road. He listens to their responses. They say the land here is very steep; the longhouse poor; vehicles and gas expensive. Going down by boat, you can float; the road will not last long; they cannot afford to use it anyway. But, I ask again, if the logging were stopped tomorrow, wouldn't they have some hardships? Again the question is translated; people talk. No, the jobs go to a few only; a lot of money to a few only. John speaks softly in English: "They are losing their land rights forever. What every native wants is tribal land recognized and inviolable." His remark is translated; a murmur of sad assent ripples through the circle and outward to women and children in the shadows against the walls.

I probe again, to hear it from their lips: What about cash crops, using the road to market? "The plantation person is never local," says the thin-lipped man. "He is someone from outside with the cash or credit to plant five thousand rubber trees. Suddenly he will come and cut what's left of our tribal land, he will have a paper saying he *owns* it." The man gestures out through the wall to the darkness, and eyes follow his hand to the green hills they know by heart. "And he will pay us low wages to pick his tea or coffee. That's what happened at Temalah." He smiles. "The lawyer who signed the affidavit for the company against Uma Bawang—he's Kenyah, from Uma Bawang, lives now in Kuching and works for the companies—as his reward he has a big tea plantation near Temalah, and Kuching points to him and says the natives benefit from logging." They all smile. I can smile no longer, but they can.

Another man, older, now speaks; the others quiet down, polite and attentive. He took part in signing the agreement with Lambir, in 1988. All the companies act alike, he says, Lambir is just a little smaller and more amateur. One day, with no public announcement, the Lambir agent and our friend the Penghulu went up to Long Palai and came back down picking up every headman in the middle Baram. "We went to the Penghulu's store in Long Akah, drank free beer and were given $20 each. The company agent said, 'My boss is new to the Baram and wants to meet everyone, wants to get to know you.' We all went down to Long Lama [five hours downriver]. We were given $75 each, drank some more beer. The next day we were taken all the way down to Miri [a full day's trip by express, at the mouth of the Baram], put up at the Park Hotel [a very fine place by any standard], given drinks, the seafood buffet and another $200 each." By the third day, they were quite impressed with Lambir management. They signed without even going home. The longhouses upriver knew only that their headmen had left in a boat.

The Lambir agreement is the one that said each longhouse could protect one hundred trees as a forest preserve if they got a paper from the government. No application letters were answered.

We break for a while. Women bring hot, sweet tea in little glasses. I sit against the wall, wearing, like Richard and John, jogging shorts and a T-shirt rolled up above the waist. The heat is always there, waiting for a stillness in the air, a slight rise in metabolism. One sip of sugared tea—I mean one sip—and in thirty seconds I am sweating all over. I put my notes beside the candle stuck on the floor, and confer with John, Richard, the thin-lipped man, and a few other leaders. I want to review the economics, the total income to one longhouse from a logging operation. In response to my questions, they patiently review the entire situation.

First, the bribes to a headman, up front and per month. These seem to vary a bit with the size and importance of the longhouse, perhaps also of the headman. At Long Anap, the headman was given $200 up front. He was promised $115 a month for the duration of operations; the assistant headman, $58; each of four committee members, $20. Those, as at Long San, were the signees of the agreement. Total cost to Samling, about $250 per month for bribes. Unreported and undisclosed, to secure signatures—what else can you say? Presents are neither secret (this was not a love affair) nor conditional.

Second, payments to owners of cultivated land crossed by roads. A typical rice paddy is about thirty square chains. At $11.50 per square, that makes a one-time payment of about $350 to each affected individual. The road is just approaching Long Anap, so no land has yet been lost, but down at Long Selatong, for example, four people were paid. Total cost to Samling, $1,400 once.

Third, payments to the tribe for logging on tribal land: $0.20 per hoppus ton. No one outside Samling management knows how much is actually taken from each longhouse's land, but since the value of a meranti log at Miri (the cost of floating it down is not enormous) is about $200 per hoppus ton, the longhouse is offered one-tenth of one percent of the value of the log, and is probably given much less. Guessing from total amount of wood extracted from the middle Baram, assuming full payment by the company, and giving each longhouse an equal share, maximum payment would be about $100 per month to a longhouse, or at Long Anap, about $1.25 per person per month. This is wildly higher than any actual payments, as the company says tonnage and value is much less.

Fourth, general goodwill payment to the longhouse: to Long Anap, it is $87 a month for the duration, or about $1.10 per person per month (Since most payments stay upriver, I am dividing payments only by the eighty people living at Long Anap, not the 800 on the tribal roll.)

Total cost to Samling, prorating the one-time costs over three to four years, is about $385 per month for the logging rights to Long Anap's land, or for the entire middle Baram, about $3,000 per month.

Income to headman, no paddy crossed: $116.35 a month, plus enough money up front for new zinc roofing.

Maximum income to an individual at Long Anap who is not headman or on the committee, who is not one of the three or four whose paddy is crossed, and who simply loses the forest (vines, wild fruits and nuts, wild meat, fish, wood for boats and houses, a way of life): $1.35 a month, for the three years that the boom lasts.

So the entire payment by Samling to natives in the middle Baram is about $3,000 a month, against $3.6 million gross; a little less than one-tenth of one percent.

I remind Richard that the company says they'll have jobs. He smiles and explains to the group how it works. The good jobs come in from outside; fellers, truckers, cowboys, and others, are Ibans or from long-houses downstream, experienced. Three or four boys from Anap may get work, and then—as happened to him—after three years the action will move on. Who wants to move to the upper Baram, or live in a logging camp apart from family and tribe? Most do not. Some do, and can make money by leaving family. This is called modernization.

Richard shakes his head. He has seen it happen at Long Bangan, but these people, his own family, can hardly believe him. Or rather, they can believe but cannot imagine what is at stake, what has been sold. They tried their best not to sell. The government in Kuching never consulted them about a timber concession, the company says they have no rights, they told their own headman not to sign. Yet it is coming, two miles away. Little Twyla is kneeling at the edge of the group. When I look at her, the baby gibbon climbs back and forth and around, in clear distress, sometimes swinging on her hair, without ever leaving her head or shoulders. What happened to his mother, I ask Richard. "We ate her."

Much later we have gone to bed, as the saying goes, though in fact the bed appears under us when we spread our sheet envelope on the floor where we sit, with eight other people sprawled about, some snoring. Light of the full moon floods through the opening in the back wall, and the frogs and cicadas come and go in waves of sound. I remember working on Native American testimonies. What struck me was that every single person who remembered the buffalo days would have preferred to have stayed in that traditional life. Men, women

—Plenty-coups, Pretty-shield—Crow, Blackfeet, Assiniboine, Sioux, Nez Perce. None wanted the white world, the modern world. As a child I was told, in my private white school, that they had suffered from disease, hunger, dirt, cold. We brought them civilization. Exactly what Chinese in Kuching, last week, had said of the Penan. Mr. Lee, editor of a Chinese newspaper in Kuching, had flown in by helicopter to Mulu Park and inspected a nearby settlement. "They lead terrible lives," Mr. Lee concluded. He wrote a government report. And here we lie in a longhouse. These people do not seem dirty, ill, unhappy. I think of Blackfeet Indians in January, on the Montana plains, the Two Medicine River iced over. In the tipi they are hunkered down together, man and wife, under layers of buffalo robes, fat with dried meat and serviceberries, the fire glowing down to embers. They are happy, happy enough, no more or less happy than me or you. They would prefer to be left alone. Or, if we really want to help, we could help them do their own developing, on their own land, in their own time.

At least the thought of the Montana plains in January has cooled me off. Outside, the cicadas buzz like model airplanes. Richard snores away. Juliette is sound asleep. Moonlight floods in through the big open hole in the back wall. The air is fresh and sweet.

CHAPTER 11 🐒
TIMBER CAMP

High above Long Anap's gardens and houses, up on the ridge, Samling's Tebanyi timber camp was going about its business.

They had just finished a record October of sixteen thousand hoppus tons extracted, and were in the middle of an ordinary month: ten thousand tons. Samling is not connected to the big Japanese trading houses and sells directly to independent plywood mills, mainly near Yokohama. The company is owned by Sarawak Chinese, and the field managers are Japanese: Mr. Sei, the regional manager for the Baram, and Mr. Fujino, manager of Tebanyi camp above Long Anap and of Jerinai camp, further north toward the Tinjar River.

On one of many trips upriver, I met Fujino in Miri and he drove me up to Tebanyi timber camp, an all day trip on the new dirt logging road from the coast to the mountains deep inside. On the way, all of it through logged areas, I had a good chance to see what happens after the forest is cut, and once there, I would be able to see the edge of the industrial world, in one camp, poised above Long Anap.

JUST OUT of Miri, our Toyota pickup is speeding up a dirt road through one of the first and largest oil palm plantations in Sarawak. We bounce for thirty minutes through rows of mature palms where twenty years ago was solid lowland forest. The plantation is making money, Fujino says, in spite of America's attacks on Malaysian palm oil. We laugh. American scientists find cholesterol in Malaysian palm oil; Malaysian scientists find it in American soybeans. We decide that who funds the labs makes a difference. Here and there among the palms in rows, workers cut the fruits full of oily seeds. At the edge of the plantation the land opens out for miles, an occasional scraggly stand of forest alternating with the total deforestation of slash and burn agriculture, which has followed the logging. Fujino says this forest was cut (by another company) at least four times in less than twenty years. No one pretends that this is a sustained yield pattern. Then the slash and burn native farmers came, and cleared what was left to plant crops. The soil is not fertile for long, so the situation now is bleak, he says, a deforested and degraded land; slash and burn agriculture is a huge problem. I observe that a well-run plantation may indeed be superior to this situation, but that the degrading began with too many logging entries, too quickly, to keep a sustained yield forest. Yes that's true, he replies, but no one ever intended to keep a forest here.

I realize we are not discussing forestry, but a frontier; they are clearing New Hampshire, Ohio, the West.

I ask if the natives working the land here are local. He says no, they are mostly Iban, who have come over from the Tinjar to homestead along the new roads. The land does not revert to the local longhouse after the timber is harvested? No, the government owns all the land, he says. After logging, it is open to settlement, and settlers follow the road. I ask if perhaps the government might have contributed to the problem, then, for if the land had remained under the control of the longhouse, those permanent native residents would have managed logging contracts more carefully in the first place, and afterwards they would have controlled the squatters. Probably, he says, but the government wants the logs out, now. The government wants the revenue.

As we begin to ascend the hills, the forest is more extensive scrub with leafy vines, and here or there a tall tree, *enkabong* (illipi nut) or *tapong,* or a steep summit with big trees untouched. We come to the Tinjar and abandon the pickup in a defunct logging clearing by the river. An old man has a contract for the ferry here, and we cross the river in his hollowed longboat. On the other side, we climb into another Samling pickup. The new bridge over the Tinjar is not quite finished. Last week it rained heavily in north Sarawak, and the high water mark on the Tinjar is twenty feet above our boat. As we cross, I ask about the Tinjar bridge, thinking of the road JICA funded on Wong's concession. Does the bridge have Japanese funding? Oh no, Fujino smiles, "United Nations." I am bemused. Why is that? "This road," he says, "will connect the Baram to Limbang, which has no direct access from Sarawak except by sea; the state needs it."

I can't believe it. "You mean this will connect to Temalah, then up to Long Seridan, then to Long Napir in Limbang?"

"Right. Miri to Limbang."

I laugh. "But that's the same road JICA funded in Limbang, for C. Itoh and James Wong, and got in trouble for. Now Kuching has talked the UN into funding it?"

"I guess so."

"For the good of the people, right?"

Fujino laughs. "I don't know what they do in Kuching."

Beyond the Tinjar there is little shifting cultivation, and the forest after "four to six entries in twenty years" looks pretty thin. The road has more gullies and washouts in these hills. Fujino explains that we are not on Samling's road yet; this one belongs to Sarawak's biggest timber company, Thien, and now that their concession is finished, no one maintains it. Samling will have to bring down some dozers before it's impassable.

The hills are steep now, and we are in them. We round a sharp mountain corner and come to a sudden stop at a huge red landslide. I am amazed. "This is usual," Fujino says. "Heavy rain." The scale of erosion is enormous. I stand beside our truck, in a line of twelve stopped vehicles (all with timber company markings), and look at a steep, naked hillside rising eight hundred feet. I don't know how many hundreds of acres have slid, smooth and bare; it looks like a red mud parking lot the size of Central Park up on edge, a ski slope dream turned into a nightmare of mud. There is no sign of where the road once traversed the slope. I wonder if our trip is over, but everyone is joking and watching six Caterpillars working the slide. One huge D9 bounces down, jetting blue diesel smoke, blade lowered and red earth spilling away in waves. He is plowing a new road diagonally down the hill, in one swath. A mound of red dirt twenty feet high rolls ahead of him. Every fifty yards he pushes it over the side and the dirt cascades hundreds of yards downhill. In twenty minutes the Cat has plowed a half mile down and across the slope. Trucks start their engines; the D9 backs up the entire distance, flattening the earth as he goes, and trucks follow immediately up the steep new road.

That frontier energy, at once attractive and frightening, a world of men and boys at violent play, takes another familiar form as five minutes later we round another corner and slam on the brakes. All the vehicles are stopped again, but this time everyone is out and running along the road, pointing downhill, laughing and shouting and throwing stones. The lead truck has hit a wild pig, a sow, and the piglet is running on the bare slope below the road, panicked, unwilling to leave the area, and finally, stunned by stones. A boy proudly holds the piglet up by the hind legs, facing my camera. It is kicking and trying to bite his hand. Others have, in three minutes, skinned, quartered and divided the sow between the lead trucks. The baby's throat is slit. They throw the carcass in the back of a pickup which is already moving off, full of boys in high spirits hanging onto the railings, laughing at their good fortune.

I think of Lewis and Clark ascending the Missouri in 1806, shooting every grizzly bear they see; and of the good Jesuit father Nicolas Point, floating down the Missouri in 1847, his men firing at animals along the bank without stopping to get the meat. Father Point was offended, even though game was plentiful, and he made them cease. These pigs, at least, will be eaten, but the energy is the same. Frontiers have fast trucks, fast boats, fast horses, and soon young men are the only thing running wild.

Three mud slides later we reach Tebanyi camp, beautifully situated on a ridge between the Baram and the Tinjar. A clearing about a half mile square descends the hill in steps. From the administration building, with its offices, manager's quarters, and rooms for guests, we look out over equipment sheds and workers' dorms to huge forested ridges receding to the horizon. To our left, the twin-peaked Mount Kalulong rises through evening mists to over five thousand feet. All about us the land is steep and green; I cannot even tell if the surrounding hills have been logged or not; Fujino says they were harvested in 1987, first and second entry only.

The next morning we go out for a pickup inspection of the coupe. A coupe is a square area within a timber concession, designed (and licensed) to be finished in one year. The coupe is divided into thirty to fifty blocks, each worked by a single team of four men: feller, cable tier ("hopman" in Sarawak, "choke setter" in America), skidder, and barker. Debarking must be done within the day at roadside, or insects get in and eat the wood. Fujino says in two days a tree can be ruined. A team finishes a block in about five months. Tebanyi camp has about fifty teams, so they could cut two coupes (100 blocks) a year but instead they reenter the same block for a second cutting.

A team is paid by the ton extracted, so when they go in, they look for the richest forest, which pays the most per day (as one might imagine, the best teams are rewarded with the richest blocks). A skidder can take out about a log an hour. After a few days in one area, they look up at the steeper, less accessible terrain; it will yield only a log every two hours. Less incentive, since they are paid by the ton. They move on.

The company comes back after first cutting and tells them to get the rest, offering a higher price per ton. Then they enter again. First and second cutting will take every marketable and accessible tree over two feet across at breast height.

The road is traversing a steep hillside; we round a sharp corner to find a big Mercedes logging truck blocking the way. We stop the pickup and get out to watch as the truck driver and his lift operator (a dozer with pincers on front, like a big forklift), assisted by the skidder's D6 dozer, nudge thirty-foot meranti logs into the pincers and up on the truck. We are at a steep bend of the road. The road cut has created a mud slide two hundred feet high above us, and three hundred feet below. At least ten acres are washing downhill, to the streams, to the fish. The two dozers take twenty minutes to load three logs onto the truck, maneuvering right up to the brink of a thirty degree mud slope falling away below them. Words are painted on the cab of the Mercedes truck, above the front windshield: "The Logging Emperor."

"That's not too bright," I say to Fujino, "given Japan's reputation in Southeast Asia."

"Not too bright," he agrees, laughing. "Those are Komatsu's decals. They put them everywhere."

We drive a little further and come to a tributary to the Tinjar. Though the road continues on the other side a few miles, we are at the end of logging on this side of the Baram. On the other side, I have already seen Lambir's concession, where Richard cut the meranti. Like the Long San bridge, this one is not finished, waiting for parts from Britain, and the river is too high to ford. That doesn't stop the Samling pickup in front of us, however, with its three teenage boys and an arc-welding unit lashed to the bed. They call a D6 operator over from the opposite side, lift the front end of the truck on the Cat's rear cable, and are pulled across the river, water pouring over the truck's rear wheels and into the door as the dozer churns through the rapids. They are having a good old time, sitting on the hood. Below us down the bank is a yellow diesel crane lying half buried on its side. Fujino laughs. It belongs to the bridge company and was parked on the road, but in the heavy rain last weekend the road slid and took the crane with it. In four days a creeper has covered one track and entered the cab. I would have guessed it had been there a year.

We turn back, and on the way meet Kenyah children coming home from school. They walk, with a few adults, along the new logging road, swinging their satchels, then disappear into the forest, toward their longhouse. Closer to camp, we come across an overturned log truck that was not there an hour before. The windshield is not broken, and no one is in the cab. The driver clearly lost control on the outside of a bend and went over with a full load of logs. Fujino thinks the truck is not badly damaged. Someone has already radioed in (every truck has a radio phone). On our way back we pass dozers clanking down the road. Fujino stops, calls in their numbers to the dispatcher, and makes decisions on placements—which blocks to cut and when depending on weather and holidays—and redirects the dozer operators. They clank off. I think of all the heavy equipment I've seen in one hour, half of it in peril, deep in Borneo. The overhead on this operation must be tremendous.

What does a bulldozer cost? I ask Fujino. He says $115,000 for the little D6 skidder, and on up to the D8 for road construction and the D9 at $350,000. Komatsu's 155A, however, is beating out the Caterpillar D9 for the same jobs at two-thirds the price. But then, the Caterpillar is manufactured in Japan in a joint venture with Mitsubishi—I had wondered why I'd seen so many. I ask about financing for the equipment. Most is Samling's or leased, he says. The Japanese trading companies used to finance quite liberally, but decided it was too risky and in the late 1980s have been pulling back from field operations. Of course, I think, they're also running out of timber in Southeast Asia.

How much does a road cost to build? About $31,000 a mile for the main road, designed to last the life of a five to fifteen year concession. And if a Komatsu dozer is $250,000, and several are necessary to keep a road open. . . . I ask who maintains the road after the concession is finished. "No one." "Doesn't the government take it over?" "No." But, I protest, the longhouses want roads and think they are getting them. After the logging, can the natives maintain the road? "The longhouse cannot possibly do it," Fujino replies. "Without this equipment, the road's no use."

Back on the balcony of the headquarters, we look at the forest, green and relatively intact after two entries. "Samling will probably cut only twice here," he says. "These people are lucky, in a way, that the forest is so poor. We're taking only one, two, three trees per acre. The Philippines had eight times the tons per acre. Here it's not worth a third and fourth cut, though in a ten-year license we might come back, or lease out cuttings to other companies. Some small ones go after the poor timber." The ridge across from us is so steep that two-thirds of it is virgin forest still. The steep terrain guarantees that no matter what the impact on accessible areas, plenty of seed plants are left of all varieties, spread about the coupe.

"I think in fifty to eighty years you can have primary forest back in hill country like this, if there are only two cuts," he says. I am surprised, but Fujino has been very straight with me, and he knows his job. The trouble is, so far, no one has stopped after two cuts. "You could have sustainable forest all around in Sarawak, if you took second and third cuts over eighty years instead of five, ten, or twenty. A meranti matures in eighty years. It's very fast growing. After that, it begins to rot."

"What I can't figure," I say, "is why they don't leave some seed stands here and there, across the country, just patches along ridges, and protected corridors along the riverbanks. You could still cut most of the timber."

"Everything in the coupe is for cutting," he says, and laughs. "The setback requirement from the river is twenty meters."

"But it would have been so easy for the government to have stands of primary forest everywhere, lowland and up."

He shrugs. "We don't make the laws."

I think of the beautifully managed forests in Japan, the neat rows of cedars not being cut, and wonder what these managers really think of their host governments.

After dinner, we are upstairs on the beige sofa in the white planked sitting room of Tebanyi timber camp, moths inside near the fluorescent lights, cicadas outside. I look back over my old notes, to give the company a chance to correct errors. According to Fujino, the timber wages I had heard in the longhouses, from Richard and from SAM, are quite accurate, from the lowest paying jobs at about $115 a month on up to Cat and skidder operators at $400 to $800 a month. The one discrepancy is truck driver pay: Richard and others had said a driver could earn up to $8,000 in an extraordinary month. Fujino and I work out the pay rate and tonnage; $8,000 seems improbable, though it depends on the distance trucked. Fujino says that he has never heard of anything above $3,500 in a month, which is high enough, and Samling drivers earn less, topping out at $2,000 in the record month of October. Let's say that truck operators in good months earn American middle-class wages, while living upriver in Borneo.

I know for a fact that fellers and skidders decide on the trees to be cut, and in the absence of company representatives or government overseers hold the forest in their hands. Some people had told me that the fellers and their crews are paid for cutting even illegal or unmarketable trees. I found this hard to believe, since the practice would waste company time and equipment, and of course such a practice would be unwise since it offers no incentive to save any part of the forest at all.

Fujino said that Samling does not pay for unmarketable trees; the grader at the riverbank rules on acceptable tonnage, and the crews and drivers are paid accordingly. Thus there is no incentive for the feller and skidder to waste their time on useless logs. However, he said, some companies *do* encourage their fellers to take anything; otherwise, "they fear the feller will leave good wood up. Don't judge it standing, they say, put it down and take a look at roadside." They are paid for all trees dropped, regardless of later rejections.

What about illegal trees, I ask. "We know only *enkabong*," Fujino says. It is up to the forestry official on the raft to identify illegal species and levy a fine. From other conversations, I know that illipi is marketable wood—just forbidden—and the implication seems to be that nothing is illegal till the whistle blows. Fujino has told me that Sarawak has almost no officers in the woods. "In Indonesia they visit operations and are quite strict." Here, the forestry officials stay in Marudi, where by law they are supposed to inspect passing rafts and personally nail a tag to each log, for certification, tax records, export counts, and so forth. A rafter told me that he has seen a forest officer at Marudi give money to a cowboy to nail a tag on every log, without once coming out to the raft. The convenience to the company is obvious. So is the opportunity for corruption. The company can pay the forest officer to look the other way. The only real screening is by the Japanese shipping agent at the mouth of the river, who buys all the trees that plywood mills can handle. These are some of the reasons that for ten miles up from its mouth, the Baram's banks are thick with abandoned logs.

Long Apu, just downstream of Long Anap, lodged a complaint last year with the forest officer about illegal felling of illipi trees on their tribal land. The officer checked and saw the tree stumps. Later the people asked what had happened and wanted to file for damages. The officer said the company (Samling) had been fined, but he would not say how much or when. Since he would not produce any document or record of the action, the longhouse could not apply to the government for damages, or bring action against Samling. In Sarawak, not once did I hear a story of the forest officer defending native interests against a timber company.

Looking at my notes, I remember that Long Anap told me they had several arguments with Fujino, and blockaded his road. He doesn't know I have been there, just a two-hour walk downhill to the river. They told me he was hot-headed and angry, impolite; but the native style is so mild, I can imagine Fujino, the professional trying to run a timber camp, pretty frustrated with local resistance.

"What about Long Anap?" I ask.

He smiles. "They blockaded us September 23, and again in November. Stopped equipment in Block 3 of the coupe. It's very complicated at Long Anap. Their headman moved over to the Tinjar about ten years ago, his replacement was a problem and they threw him out, and they invited the old headman back. He's been there two years. He's the one who signed the agreement, but they're claiming the agreement wasn't signed by a legitimate headman."

"What do they want?"

He laughs this time. "They want their entire tribal land protected. That's most of the '91 coupe. 2,000 hectares."

"Do they want more money, or no logging at all?"

He laughs again and shakes his head. "Long Anap is a little different. They've made it very clear that they don't want money and won't discuss terms. They just want us to stay out. We can do nothing. I called in the District Police, but there was no blockade at that time. We'll just see."

I sit back and smile. That's something, anyway. But I remember the piglet, kicking and squealing as it was hoisted by its hind legs, and I wonder what these people can do against fifty teams, 250 bulldozers, over $2 million a month from one timber camp. We have our last cup of tea, and, as the night rains begin, go off to bed beneath the pounding of water on zinc.

In bed, alone in the dormitory, with the luxury of a light since their generator never stops, I pull out the Samling ownership papers which I had unwisely and inadvertently brought with me. The papers, public records, were legally obtained through a lawyer in Kuala Lumpur, but a police search—which could happen to a bearded man with a backpack at any town, any dock, any airport—coming on these papers . . . let's say you'd spend days answering questions, everything you had would be inspected, and very possibly your notebooks would be permanently mislaid, in a safe with Bruno Manser's.

I am trying to think realistically about Long Anap. What are they up against? Is it all over? I know that Samling is a big company, and experienced, perhaps hardened, to upriver cutting and confrontation with natives. Samling was the contractor for the bridge that opened up the Tutoh to the Limbang border, the area that had been the stronghold of Manser and the nomadic Penan. Samling's bridge was one of four burned mysteriously in 1987. I know Fujino had been the agent in Limbang for James Wong's wood at one time, and that Mr. Sei, the Samling regional manager, had also worked in Limbang (probably with Wong). I never could find out if they were associated with C. Itoh, the Japanese firm that had built Wong's road, but it is highly likely. None of this is surprising, since one would expect such professional men to be avidly sought, and one would expect C. Itoh, when it pulled out of Limbang, to place its experienced field officers somewhere in Sarawak. The point is that Samling by now knows exactly what it's doing, and its officers have faced plenty of native opposition before. These men know their options.

So who is Samling? The ownership papers tell me that the directors, managers, and shareholders (four) are Chinese, live in Miri, and are mainly from two families (maybe related). I see that they have no trouble acquiring operating capital: loans in October 1990 of $15 million for "machinery and equipment," and before that in June of almost $10 million, both from Malaysian banks. Then loans in 1989 of almost $8 million and another for almost $6 million, from a third and fourth bank. Then in 1988, two loans from Citibank—that's interesting; I remember the report that the Hong Kong head of Citibank flew in to

Long Akah with his wife in twin helicopters—of almost $6 million and almost $4 million. And so on. I am no economist, but I can reach two conclusions: Long Anap is playing against a team from the big leagues, and Samling's operations seem to be expanding (in three years, loans increased from $10 million to $14 million to $25 million). I could draw that curve, though I'm afraid it might veer off to Siberia or Brazil. If this is sustained yield, I'm the new public relations manager of Mitsubishi.

I'm having so much fun now, watching moths like tennis rackets whacking at the ball of my light bulb, that I decide to look at timber industry propaganda I have picked up from Samling and others. The oft-published Japanese "forest specialist," Ichisaburo Morita, is one of my favorites. He tells us that "the tropical forestry is sustainable industry in Sarawak," and that "logging is not the major cause of tropical deforestation." This last may be true worldwide, but is completely irrelevant to Sarawak—and this article is headed "Logging Sanctions Only Hurt Sarawak Natives." Morita ends with inside information: "My recent visit to Sarawak confirms no people want to live in poverty forever." An astute observation. For a long time maybe, but not forever. "The alternative is to assist natives to develop other industries of high values or alternative sources of income, such as tourism."

Ah yes, the call to modernize the Penan against their wishes but in their best interests, while even pro-logging Sarawakians admit all the "development capital" is leaving the country for private, offshore bank accounts. And what about tourism: the Long Anap Hilton in cutover jungle; raft trips on brown, fishless rivers—"Hey Martha! Isn't that a log?"—and five-day treks in oil palm plantations, two nights in a longhouse, curio shops with indigenous arts and people in traditional dress. The timber industry is helping natives to enter the new age.

An Iban in Kuching told me that his longhouse got onto the tour list, and foreigners left quite a bit of money there, for baskets and blowpipes and such. But the Iban got tired of changing into sarongs and putting on their straw hats on Friday afternoons, and when they decided to just keep on their T-shirts and Adidas shorts, their longhouse was dropped from the tour. Local color must be carefully managed for sustained yield.

I am getting sarcastic, I think to myself, because I have read all this nonsense before, and it has nothing to do with reality. Even James

Wong had said that the present cut is sixteen million board feet a year, and sustained yield would be eight. And even the International Tropical Timber Association (ITTO), the industry's own mouthpiece in Yokohama, recommended a 30 percent reduction in annual cut. So here's Wong saying 50 percent reduction, the ITTO 30 percent, and the timber camp proving to me that four cuts within twenty years—the norm—destroys the rainforest. Meanwhile Japanese newspapers feature Morita's claim that "forestry is sustainable industry in Sarawak." If the timber industry wanted to be honest and accurate, they could say, "When the primary forest is gone, we can sustain 20 percent of the present cut by taking different, less valuable species from planted tree plantations. The rainforest and native economy, however, will be gone, and planting has not yet begun."

As I lie in bed, I realize that the timber industry propaganda, not the environmentalists, has convinced me: the logging is indefensible. If they had a leg to stand on, they would be able to present a cogent argument. I have read all Wong's replies, heard timber people in friendly and confidential, private conversation, and read the ITTO reports. SAM did not convince me that the logging is bad; the loggers did.

Morita, the Japanese forester quoted in the *Japan Times,* is serving the industry organization, ITTO, funded by Japan, and he is passing on what I know to be lies: "The rural natives are benefitting from the timber industry, and they are not, with a few exceptions, against logging, the ITTO report said." Strange, my own sample ran about 300 to 2 in the other direction, Long Anap seems less than pleased at the advent of bulldozers, and in October, Harrison Ngau won in a landslide on an antilogging platform, while opposed by every headman in the Baram. The truth is, I began this project expecting to find a much more complicated problem of modernization, gray areas, "two sides to every question." I have done my research. It is a story of the few versus the many, a story of greed.

It is disappointing in a way. I had thought the timber industry and the ministers might be more interesting.

Later, in Tokyo, I heard reports of the native opposition's delegation to Yokohama. Our friend Jewin Lihan, Jok Jau, Thomas Jalong, lawyer Baru Bian, and the others went to Tokyo and Yokohama to give unsolicited testimony at the ITTO. I remembered Jewin Lihan at the dirt road up the Tutoh, when he stood in white shirt, holding his shoes and a satchel, watching my family come up with Richard from the slippery log over the stream, out of the jungle. That was the beginning of his trip to Yokohama. The ITTO supported the Sarawak timber industry and refused to consider native land rights. The native presentations, however, arranged by Yoichi Kuroda and JATAN, were covered by the Japanese press. I saw copies of *Jump* and *Friday,* weekly Tokyo magazines, with pictures of our friends from the Baram testifying and explaining their plight.

Jewin and Jok had never before been out of Malaysia. In Tokyo, I asked JATAN how they had reacted. First, I was told, they were absolutely astounded that you had to pay money for everything—not just food, but laundry, an apartment, the subway or cab across town. And the waste: Jewin and Jok stood in front of their *ryokan* (hotel) in Yokohama, on garbage day, staring at chairs, tables, and a sofa piled on the garbage cans. They looked new, but they were being thrown away. Then Yoichi Kuroda told them the furniture was made of meranti wood and meranti plywood. They were amazed and angry.

Kuroda told me of Jewin at Nara, the famous temple city. They went first to the park with the tame deer. "You go see the temples," Jewin said, "I'll stay here." He sat there the rest of the day, on the grass in the midst of hundreds of deer, talking to them.

I'm afraid all these meanderings have quite put me to sleep, but I recall, as I doze off to drops of rain against the background of a generator's hum, that Fujino has convinced me there is time left, in this hill country, to save something—first and second cutting is not the end. I remember also Harrison Ngau smiling when I asked if land rights would be worth recovering after the logging. "The natives want their land back," he said, "every inch, no matter what has been done to it. They will want it back." The soft wings of moths, and the shrill cicadas, like some yin and yang of this fecund land, sing me off to sleep in Tebanyi timber camp.

CHAPTER 12 ❧
MEETING AT LONG MOH

Once SAM discovered that the upper Baram meeting with Samling would be on November 27 at Long Moh, they made plans to send a longboat from Marudi, picking up support along the way, to advise the natives of their options. The expedition would be led by Joseph Wang Tingang, a politically experienced native leader in the Baram, and would include Richard, John, and several others I knew. I would go with them; Juliette, however, was not feeling well, and needed to return to Kuching to finish her article on Penan dance for the *Sarawak Museum Journal*. She and I went down together, then I rejoined SAM in Marudi. The trip up would be fast—one continuous journey from Marudi to the top of the Baram. I would see the country from town to far beyond logging in one sweep, and would travel in the company of men who knew the situation well. Joseph's impeccable English and knowledge of Sarawak politics would soon prove a great help.

Saturday, November 24, 1990
Joseph and I walk down the white, dusty street of Marudi, past equally white stores already in full sun at 6:45 a.m.—Wong's Motor, Chop Kiat Hin ("Husqvarna"), Chop Keng Soon General Store, Las Vegas Pub and across the street, Great Wall Dentist. A packed express boat, crates on the roof, kung fu videos. Upriver again, past Joseph's home at Long Ikang, past Harrison Ngau's small longhouse of Long Kesseh, past Uma Bawang where Ngau's celebration was held and Jok Jau's fruit trees were bulldozed—he has just returned from Yokohama —past the various timber camps (three hundred new coils of logging cable and four new Komatsu dozers on the mud beach at one) and raft after raft of logs nudged down by cowboys, fifty logs in a raft, another thirty acres of forest gone.

Back at Long San, we find that the Tattan owner went upriver with a boatload of beer and came back yesterday. The Penghulu went up twice and goes again tomorrow—all for the meeting, says John. Why are the Tattan owner and Penghulu going up there? I ask. "Spread company money and beer, tell them they have to sign anyway, say SAM is sabotaging development and trying to split up longhouses. Same thing the government says. They're paid for it."

The next morning, in a steel launch for which we'll each pay $60 gas money, our first stop is Long Selatong, where they blockaded a few months ago after Samling cut their *enkabong* trees. Our room at one end of the shaded, flowered longhouse is forty feet square, open above to the high roof, light and airy, with overlapped strips of reddish hardwood for a wall. The strips and floorboards were planed at the mission at Long San; the roof took one hundred sheets of zinc, bought with rubber income. The rest was free for the taking, and will be until first cutting, within this year. We are at one end of the longhouse, with those opposed to the logging. The headman is not here; last week he, too, went up to Long Moh in the company boat, with the Penghulu, the Tattan owner, and the beer.

Women serve rice to a dozen men and women in a circle on rattan mats, then sit and join in the discussion as it becomes more animated and political. Men and women begin to speak more vigorously, though always in turns, in Kenyah.

After lunch, we add two men from Selatong to our boat and two more from Selatong Ulu across the river, and make it up to Long Apu for the night. After dinner the kerosene lantern is lit, a Fuji clock on the wall plays—I swear—"Old Black Joe," and that night's local meeting begins. Joseph holds forth. Richard translates; they are discussing what they should tell the upper Baram about Samling, logging, and timber agreements. Here I learn that although the official meeting with Mr. Sei is day after tomorrow, the natives will meet privately in Long Moh tomorrow night to hammer out their stand. That is when the real work will be done. All the natives, pro and con, want if possible to present a united front.

Listening and watching that night, I am struck once again by the gentility of these people. It is hard to describe but instantly sensed in person, the bright eyes behind a shy manner, the soft handshake, the arm extended slowly as if you might run away, like a wild animal; the speech never loud or aggressive, even when animated; the children almost never whining or crying, the parents almost always kind. Once, at a big longhouse way downstream, I saw a woman angry at a ten-year-old boy, raising her voice and slapping him. People were very embarrassed, looking away, and I realized that it had been over a year—when I left America—since I had seen such harsh discipline of a child. When Bruno Manser entered the room in Tokyo, after seven years with the Penan, he seemed to glide in, to fade into a seat. Our daughter, Mary Catherine, who has lived in America, Europe, and Japan, said she had just seen Bruno use three gestures of head and hand she'd never seen before, all soft and self-effacing.

The next morning we pick up more people at Long Pelutan and stop for a visit with our friends at Long Anap, then up to Long Palai at the top of the middle Baram. There a funeral is in progress and the longhouse may be closed, but we are allowed in past the shaded boat-building flat and the giant sago palms, across the green soccer field to the big house, 150 yards long. We visit. At one end of the longhouse is the skin of a giant anteater. I find a British Enfield rifle on the wall—left over from Tom Harrisson's campaign against the Japanese, says the ancient owner.

Then upriver again, Richard at the helm, our sixty-foot boat with twin Suzuki 40s packed almost full, roaring through rapids and around the big bend, into the upper Baram. Now we are beyond not only all logging but all roads, and it will be an hour to the next longhouse. The hills are close together and steeper than any forested slopes I have ever seen, the jungle canopy rising up in waves of green. The river is pinched and very fast, swirling with brown boils and pressure waves; we run as close to the bank as possible, under giant trees, in between trailing lianas, the dark, virgin forest gliding by. One tree per acre. For one tree per acre, this will all change.

It is a good time to huddle with Joseph Wang Tingang, up near the bow but still under the boat's metal awning, out of the sun, and as far away from the roaring outboards as possible. We sit on wooden planks laid across the boat's bottom. Now and then Joseph and I have to shield our papers from spray as we crash through waves. Joseph is hunched over; long and skinny with thick glasses, and he looks at once native, Chinese, and American. He speaks quickly in a number of languages, including my own. From Long Ikang in the Baram just above Marudi he is district vice-chairman of a political party (SUPP) and heir to family interests up to Long San, where his grandfather gave over twenty acres of land to the Dutch Catholic mission and its school. His father, less devoted to the support of Christianity, was not very happy about that. He laughs. Joseph knows every headman and Penghulu up the Baram and is related to many of them, and is the only member of his establishment family who supported Harrison Ngau in October's election. Thus he lives in both the tribal world and the world of democratic dissent. Those two traditions are not easily reconciled, he says. "If I think of relationships, I cannot move at all."

Joseph had a ministry post as political secretary to the Institute of Science and Technology in Kuala Lumpur, and returned seven years ago to the Baram, once standing for and losing the seat that Harrison has just won. He's Kenyah, with numerous rubber and fruit lands, a huge, separate house in Long Ikang, and other doors in Long Akah and Long San. He is intelligent, articulate, and respected. He just happened to be in Marudi when John called with news of the Long Moh meeting date, and the SAM office collared him to help.

Joseph and I have a chance to mix and match our statistics, from various sources. We want to put together what the company gets from the forest, and what the natives get from the company. We'll take the middle Baram as example, seven longhouses. The timber coupes do not coincide with longhouse territories; in fact, the government grids pay no attention at all to natives, so cutting is not on one group's land or another's. Still, seven longhouses at two thousand hectares each constitute a bigger district than the biggest coupe, so we'll pretend that all the middle Baram operations in one year are along the river, within middle Baram longhouse lands.

Joseph has figured the price of meranti FOB Miri at \$223 per hoppus ton, and operating expenses at \$147 a ton, including survey, road construction, felling, barking, tractors, trucking, rafting (wages, fuel, depreciation included), export and excise duties, administration (\$35 a ton), and payment to the Government Native Rehabilitation fund (\$10 a ton, results invisible). That leaves a profit of about \$80 a ton. Fujino would not discuss profit margins; he confirmed some of Joseph's expenses and figures as roughly accurate for meranti, although about half their harvest is mixed wood, less valuable, and this calculation also leaves out the cost of kickback to get the license, which in the halls of government can be high indeed.

In the middle Baram, Samling currently has only the two timber camps operating, Tebanyi and Jerinai, each producing about 10,000 tons a month of marketable timber. When the bridge at Long San is finished and the east side of the river can be worked, a third camp will add another 10,000 tons. Let's take the current month's production of 20,000 tons a month at $200 a ton gross (remember, such production could be achieved on about one-third of the middle Baram native land). That's $4 million a month gross; divided by seven longhouses, almost $600,000 a month each. Let's take Joseph's figure of $80 a ton profit, and Fujino's word on mixed wood prices, allow for some dozers rolling off hillsides and trucks rolling off curves, and cut the profit in half: $40 a ton makes $800,000 a month profit for the middle Baram or over $100,000 profit per month per longhouse.

That, until a company shares its books with us, seems to me a reasonable estimate of what a company is making from native land, on first and second cutting. Over the years, by the time third and fourth cuttings are finished, it will work out to this and more. The natives are compensated per ton at one-tenth of one percent ($0.20 per ton). As we have seen, if everything is paid off as promised, every damaged fruit tree and garden compensated, even bribes included (though this money goes to a few only) and tonnage honestly counted, the maximum conceivable return to natives is about one percent of value taken, or about 2.5 percent of profit. Every longhouse in the Baram has said no even to this paper arrangement. Joseph and I may well be overestimating company profit, but we certainly know well the scale of operations—company gross—and the scale of payments to natives.

We cruise by Long Jeeh. With about forty buildings over as many acres, it is almost a village of longhouses. We pass lusher hillside rice than I have seen, and through the twisting rapids of an ever narrower river, now and then touching a gravel bar or log. At one break on a stone beach we take a group photo while Richard and friend, "stone age savages" says Joseph, bang a propeller back into shape with a rock.

In midafternoon we pull into Long Salaan, above Long Moh, to gather more friends. Long Salaan is near the top of the navigable Baram, just a few miles below the highland plateau. It is one of the prettiest settlements in Sarawak, with several rows of longhouses and outbuildings planted with hibiscus, bougainvillea, orchids, several shrub varieties of the familiar coleus, and a dozen species I cannot name. There, on a porch with the usual sweet tea, I find just how remote we are from SAM and Marudi. Tomorrow's meeting was scheduled for earlier in November and was postponed twice; it is not a December meeting suddenly moved up. Every longhouse has a SAM contact or friend, yet no one has gotten the word back to Marudi. Mail service doesn't exist, it can be months before someone goes down ($300 round trip), and the only phone is the government radio phone at Long Moh. With the government agent there listening, a call would go to the government agent in Marudi, who would put the caller through—through, that is, to SAM, an organization the government has labeled subversive, and whose leader has been jailed. The caller would be marked, you can count on that. It is also possible, some say, that SAM officers inside the beltway of Marudi forget their constituents upstream. Be that as it may, the door next to the headman's has a Harrison Ngau poster, and an antilogging cartoon in Bahasa, used in the campaign. Joseph drew it.

At Long Salaan we meet up with four Penan who have come down from Ba Muboi and Ba Ajeng, Penan settlements up the Salaan branch of the Baram, a day from Long Salaan. That is deep in the interior, and it is also the area the east road will now approach. For this big political meeting, they have changed out of traditional Penan loincloths into slacks, shorts, and dress shirts. Then, as we are shoving our launch back into the river, having added three more Kenyah and the four Penan, a company speedboat roars up with a hardhat driver and the ample frame of the middle Baram Penghulu. He seems surprised to see us. Especially to see me. "They didn't know we were coming," says Joseph. "What does that mean?" I ask. "It means we have more friends than they do."

Down to Long Moh, a half hour away. A hundred yards up a slow tributary, Long Moh has an easy landing. Two black goats are tied in the tall grass on the bank, and chickens peck around their legs. On the river, a boat is being poled upstream and as we unload, two boys, about six and ten years old, jump in a dugout with two paddles, cross the stream, and work up the opposite bank under the trees, handling the canoe superbly. The young bowman puts down his paddle, stands up, and readies his weighted throw net. We walk up the bank, away from the river, past the sign "Selamat Datang Ke" (Welcome to Long Moh).

We have had our tea, some have bathed in the river, and we are relaxing on the veranda when another speedboat pulls in. Out steps a man in a beige safari suit with an aluminum attaché case and a hefty bag of gifts. It is Mr. Sei. He looks at our launch and has angry words in Bahasa with the boys and Richard at the dock. Later Richard tells with great glee what he said: "Whose boat is this? Who sent these people? Who paid? They were not invited." The boys did not reply. With all respect to Mr. Sei, I find the proprietary attitude interesting. The selling of the Baram is to be done without benefit of counsel.

Mr. Sei checks into the headman's quarters. We, as usual, are several doors and in this case a separate porch away.

As the shadows lengthen and the day begins to cool, I am seated on the veranda with La, the Penan from Ba Muboi, and his friends. As usual, children are gathered at the edge of that indefinable space that is us; they run away at the slightest glance. One beautiful little girl, about eight years old, begins the woman's dance whenever my back is turned, a slow, graceful twisting with one hand in the air; when I look, she stops and studies her feet. Gradually they lose self-consciousness, however, listening to La's stories.

La should be cast in this movie: his strong dark face, high cheekbones, and bright, flashing eyes are set off by a white shirt, open at the neck. He has a short, bowl haircut and a charming, ironic smile. He tells the assembly on the porch, who clearly find him hilarious, the full story of confronting the surveyor from Samling who professed he had nothing to do with a road, and who was told there was no logging agreement signed and he should leave or risk being taken for a pig. Then he tells us more.

Mr. Sei came to the Penan at Ba Muboi three times, first in July 1989 with the middle Baram Penghulu, bringing coffee, sugar, and rice. Those settlements are back a day's travel off the Baram, directly in the line of the crucial road to the highlands, which the bridge at Long San will serve. "Why are you bringing gifts?" they asked Mr. Sei. "Because you are poor," was the answer. But the Penan knew perfectly well what he wanted: their forest. The food was not accepted. Mr. Sei came again in August 1990, after his surveyor had been chased out, and again in September, this time with food and candy and five Stihl 1007 chainsaws. For the third time the gifts were refused, says La. The children, in front of Mr. Sei, in the clearing between the huts on stilts, poured the candy on the ground.

The others listening, Kenyah from downstream, shake their heads. "The Penan are tough," says one. "And they stick together." The Long Salaan people smile; they have heard this, and they are proud. Five chainsaws. I don't know what the annual cash income of the entire Ba Muboi community is, but it may well be less than five chainsaws. As the Kenyahs shake their heads, I find myself thinking of the spirit of Crazy Horse.

After dinner on our host's floor I push back from the banana leaf, lean against my pack, and take out the candy. I have brought two bags from Marudi, as gifts for children and adults, to last a week or more. I open them for the first time. One bag has little peanut-molasses bars, each about one-third the size of an American candy bar, individually wrapped. The other bag has big chocolate bars. La is sitting next to me; I hold the open bag to him. He takes a peanut bar. I set the bag on the floor. He takes another. I offer him a chocolate bar. He takes two. He speaks excitedly to his three Penan friends. They slide over from their conversations. Two of them, whispers John listening in, have never had candy before. They have taken off their shirts, and wear shorts only. All are young and strong, a bronze color, with black hair. They are eating. A pile of discarded wrappers is amassing on the floor. Seeing a disaster in progress, or perhaps a Guinness record, I count the peanut bars—they were packed in twelve rows of five. I remember I bought twenty chocolate bars. I think of withdrawing the packages— my week's gifts for children—but the Penan are jabbering so excitedly and tearing off wrappers so fast I don't have the heart. I check my watch instead. In twelve minutes the eighty candy bars are gone, two to Kenyahs, one to me, and seventy-seven to four Penan. I remember stories of Northern Plains Indians trading for sugar, and sitting down to eat a five-pound sack on the spot. I seem to be at a similar margin of cultures; the Penan are certainly not sick; they are licking their fingers and grinning. One of them picks up the wrappers. La asks shyly if he can have something, apparently something important—Richard is translating—the two plastic bags. They are delighted at the empty bags, and thank me profusely.

It is dark, as always at seven o'clock, and I am left alone after dinner, a very unusual situation in a longhouse, especially for a visitor. The men from our boat are spread out in different caucuses, talking to friends and relatives from each longhouse about tonight's meeting.

Women and children come and go, or sit in other areas of the bare floor, but they speak little English (or Bahasa) and are shy, especially when I am not accompanied by Juliette.

Outside, down the length of the longhouse porch, lamps are lit and groups of men—five, ten, twenty—sit in earnest, quiet circles. The next longhouse over, ten yards away, is the same. Soon the generator comes on and fluorescent tubes flicker. The kerosene lamps are extinguished, and men begin to gather for the general meeting.

This is the meeting of the natives among themselves, to discuss what their response will be tomorrow when Mr. Sei makes his proposal. The proposal will be no surprise; everyone knows it will be the same as the middle Baram agreement, for the middle Baram Penghulu, the Tattan owner, and Mr. Sei have been preparing the ground for weeks. But the response is in doubt. Tonight they will come to their decision, trying to present a united front either way.

I ask Joseph what will happen. He isn't sure, but the opposition to an agreement is substantial; however, the company knows that, and has been very active. It will be a real battle. Already houses are badly split. Headmen (and therefore relatives or influential friends) have accepted many unofficial gifts, and even purchased loyalties are taken very seriously upriver. On looking at the voting results, longhouse by longhouse, for Harrison Ngau's election, I noticed Long Palai overwhelmingly in favor of the third runner, who won nowhere else. "Oh, he held a pig feast there for a week," I was told. Since Ngau had neither money nor inclination for such electioneering, his victory was the more impressive. The fact remains, however, that allegiance can be bought.

On the other hand, many people are angry and Joseph says there is much wider general knowledge of how the logging operates than there was at the time of the middle Baram agreements. The difference, former ministerial secretary Joseph reminds me, has already forced a crucial procedural change: headmen were not taken away to a secret meeting; for the first time many will participate in or witness the discussions, at least ten people from each of the seven upper Baram longhouses. Joseph thinks most of the longhouses are ready to take a strong stand. The problems are the headmen, and despair—the conviction of the older and more conservative that they have absolutely no choice, that this is their one chance to accept company gifts.

At ten o'clock at night they beat the big drum, six feet long, hanging on the main longhouse porch, and the meeting convenes over on our verandah. To my surprise, headmen have told Joseph that I can sit in; after all, I am not a party to either side and will understand little of the debate. They also seem to appreciate someone from far away being concerned about their situation. Every woman and many of the children are sitting or standing around too. It is a town or rather county meeting, full of reunions, greetings, and chattering circles, while above them big bugs are banging into fluorescent tubes. Mr. Sei, however, is not present; he remains at the other longhouse, and the conversation is in Kenyah, which, even if he could hear it, he would not understand.

An unofficious opening: an important-looking man in a central position nods to another man and he stands and gives an informal review of issues, then announces that each headman will be asked to summarize his longhouse's thinking to date, followed by a general discussion. For the next two hours they stand up one by one, as different in style as any group of politicians, the man from Long Salaan surely Jewish with his wry smile, rimless glasses, and dry wit bringing general hilarity, a dramatic young man from Lio Matoh, strong and expressive, the old Penghulu from Long Moh, feathered headdress, tattoos, and extended earlobes. Then, to my surprise, others are allowed or invited to speak before the headmen have finished, and Joseph is speaking for a long time, perhaps thirty minutes, as I watch reactions. At first I fear his professional style and wit will not hit home, but after a while the important man and several headmen are nodding with his points. He receives strong applause.

The important-looking man is the URA, or Upper River Agent of the Baram. This government post is here taking precedence over the local Penghulu (the agent happens to be strong and widely respected). The aging upper Baram Penghulu, from Long Moh, is a good man but conservative, not likely to join in revolt on any issue. The upper Baram Penghulu is now speaking again, and Joseph answers, and they begin to debate. Soon it is obvious that these two are the spokesmen for the opposing sides, and they have locked horns. The audience of two or three hundred hangs on the exchange, and I can tell that at least all the issues are getting to the table — or the porch. Joseph is standing, holding up both hands and saying "NO, NO" in English; I hear the words "America" and "Japan." He is telling the audience to say "no" to exports, "no" to the timber trade until they regain some control of their lands. The Penghulu disagrees, but I can see the argument, though heated, is not at all hostile, and is always civil. The exchange ends in a sort of recess, fifteen minutes of highly excited hubbub.

Victor, a thoughtful Kenyah church deacon from Long San, introduces each of the next series of speakers, including remaining headmen. One of them had looked like trouble from the start; I call him Mr. Slick, the one with a blue and red Hawaiian shirt, shifty eyes, sleek hair, and something indefinably decadent about his expression. His tattoos, I privately suspect, read REALPOLITIK in Kenyah. He speaks long and smoothly, to light applause. Then I am told he is the headman at Long Tungan, the one upper longhouse which Joseph said was weak.

Victor stands again, hands behind his back, good dynamics and pacing—Dutch Catholic training on top of a native oratorical tradition—and during one of his homilies I sidle off to bed. It is almost 2 a.m., with no end in sight, and since bed is against the wall directly behind the URA, my head will be six inches from the meeting and I won't miss a word.

In my sheet envelope, clothes bag for a pillow, I am about to tune out the speeches when John comes in; they are arguing terms now, and it looks good, he says; they are naming figures Samling won't accept. However, John's report scares me, for I know that what seems like an impossible request to the natives may be nothing to the company; after all, if one-tenth of one percent of costs is spent on native compensation, Samling could triple its offer and add little to its expenses. I fear the natives playing ball with Mr. Sei at all. Joseph is possibly the only man on that porch who knows industrial economies, the scale of cheap and dear. "Will the headmen sign?" I ask; John isn't sure. I put my head down. There is nothing I can do about the upper Baram; at least they are debating their future, whether they can control it or not, in a full and well-informed conversation. I wish them luck. Sinking to sleep, I think of the Penan, eating my candy, and of their children, who had never tasted candy, pouring it on the ground.

AT DAWN, roosters are crowing and the general meeting is over, although six men are still sitting in a corner of our room, talking issues amid ten or twelve sleepers. Joseph, of course, is one of the talkers. I rise to one elbow; he looks over. "Well," I ask, "when will the upper Baram be cut?"

He smiles. "Not tomorrow. Most will not sign, maybe one or two, which does the company no good. Each longhouse stood up and said if the headman signs, he's out. And all will sign a letter to the company which says that individual headman or committee signatures on a document, without the approval of the longhouse, will not be recognized. In other words," he looks over his horn-rimmed glasses and smiles at the American professor, "the headman must have the advice *and consent* of the longhouse."

"Tell me Joseph," I say, sitting up. "Will it make any difference? Was this an important meeting?"

"Yes. Unprecedented. Never before has the initial agreement been rejected by the headmen. This meeting, plus Harrison's election, is a real step toward democracy in the Baram. If power is going away from the headmen, it must go somewhere else. Last night it went to the longhouse communities. This is good. But there are problems too. The Penan were angry; they were not invited by the company and not fully included by the Kenyah community; they spoke up. Some divisions there. Long Tungan [Mr. Slick] argued for signing."

"On what grounds?"

"He said if the middle Baram signed for so much, it's not fair for the upper Baram to ask for more."

"Are they asking for more, or saying no?"

"They will make a counterproposal for more compensation."

"Are you sure the company won't accept?"

"They can't. The terms are substantially different, and they might be forced to renegotiate every agreement downstream."

"What are the terms?"

"The longhouses are asking twenty times as much per hoppus ton general payment, five times as much per square chain of customary land crossed [I am beginning to fear this could be accepted], but," Joseph smiles, "not as a one-time payment; as a rental fee every month."

"You mean instead of twenty ringgit per square chain . . ."

"They want 100 ringgit per square chain per month."

"That's different."

"In principle as well as price. For a log pond too, at the riverbank, they want a monthly rental fee instead of outright compensation, so a continuing native right to the land is asserted. Then they want fifteen times as much per coffee tree destroyed, six times as much per rubber tree, twice as much per durian, and so on. Also vegetables for the camps have to be bought locally, and provision shops should be owned by the nearest longhouse, not the Chinese. The company can't accept."

I ask Richard's father what he thinks of the meeting. He agrees that it was very important and a real victory, but he's worried about new developments. Joseph translates.

"The URA is being given a lot of power. Last night he was elected chairman of the committee on upper Baram timber. He decides to call or not to call another meeting. He's a good man, from Long Moh. His wife's from Lio Matoh and they live there; he runs a provision shop. But URA is a government position, and he's suddenly replacing the headmen and Penghulu. He's just one man. The company, and maybe Kuching, will be after him hard, with a lot of money. How long will he hold out? And what about a different URA? We'll see."

I ask Richard, now awake, what he thinks. He sits up, looks at all of us and around the room. "Bill," he says, "I don't know what is going to happen in this country."

After breakfast, I sit on the porch with John, ten or fifteen others, and the Penan. La is carving on his blowpipe. I have never seen anything like it, and neither have the Kenyah or his Penan friends. It is a short pipe, two feet long, for the tourist trade, of beautiful red *bilion* wood. There are designs along the side, and a sight on top, and beneath, a pistol grip stock and other curly carvings sticking down. The whole thing looks like an AK 47 that came to life and grew tendrils where the trigger and ammo clip should be. He has clearly seen pictures of stubby automatic weapons, and has blended that shape with the traditional flowing, curling motifs of the jungle and native art — all in a blowpipe, the weapon he knows. It is a wonderful work of art. The carving is fine, in taste and detail.

As he works the hard wood with his knife, La tells stories, to everyone's amusement on the porch. He and his friends are selling baskets and rugs, and, he hopes, the blowpipe, if he can finish it. How much can he get? Over twenty dollars to a trader, with the intricate carving on the hard red wood. In his world, that is an enormous sum.

After touring the settlement I find myself alone with the Penan on the porch and two Kenyah; one speaks a little English. Others are doing their last-minute caucusing for the meeting with Mr. Sei.

The Penan and I discuss their situation, the logging downstream, the shelved Penan Biosphere Preserve, Bruno Manser—who, though he lived far away on the Tutoh, is well known and respected among all these Penan. I relay to them the international attention to their plight, and the realities of Japanese market demands. They have tears in their eyes when I describe throwaway plywood forms, two hundred thousand of them at the new Shinjuku building in Tokyo. They understand very well where their forest is going. Soon they will hear news more directly; Jewin Lihan, back from Yokohama, will be reporting to Penan on the Tutoh, and word will travel fast. They know who Jewin is but have never met him.

After a few minutes of silence, La looks up from his carving and says to me, through our Kenyah friend:

"Couldn't you rule us again?"

"What?"

"You're English aren't you?"

"No, I'm American."

He hardly pauses. "Couldn't you take us over?"

He has me there: Granada, Panama, Kuwait, the ghost of Teddy Roosevelt, oil fields at Miri . . . all run through my mind.

"No," I say finally, "You're on your own."

He looks down without reply and with visible sadness to his blowpipe and knife.

Our interpreter leaves; we sit carving, taking notes, reading, and talking. Then I remember I have other possible presents. Do they have a "cassette tape" recorder at Ba Muboi? With gestures, they recognize the English words. Do I have "batteries"? they ask. They have no generator. Only AA, I say. They need D. I go into the door, rummage through my pack, and emerge with three presents for the Penan: a shiny red apple (from Australia to Marudi), a bulb of garlic, and a cassette tape, "The Magic of Mozart," which includes the allegro from Piano Concerto No. 21 (the theme from *Elvira Madigan*) and the overture to *The Marriage of Figaro*. I hand over the three tiny items, each of which seems to be equally amazing and appreciated. I indicate that the apple is to eat; they have never had one; La makes an incision with a knife, looks at the juice on the blade and licks a drop, tentatively; he looks amazed; he inserts the knife again and then passes the open blade; they all lick the drops, and smile and stare at each other.

La cuts the apple carefully into equal sections—core included—and gives them to his friends; they suck on each section before eating, then crunch down. The apple is gone. They smile. It is hard to imagine that Mozart could bring more joy.

The meeting with Mr. Sei and the upper Baram longhouses has started on the main porch. Joseph and most of our crew return; along with the Penan, we are exiles from this official function, but Joseph assures me there will be no surprises. A runner comes over; Mr. Sei wants to reopen negotiations right there on the porch; Joseph counsels no, do all negotiation in writing and by native caucus only; don't let the headmen bargain face to face with Sei. We wait, sipping tea, the disenfranchised: me from across the sea, Joseph from downstream, and the Penan, who have no rights because they live in the forest, do not plant rice, leave the land alone. In an hour the word comes back— the upper Baram headmen have held fast. They have presented their proposal, and will negotiate no further this day.

I had been hoping for a wild Borneo party that night—all the other books on Borneo tell of wild parties—but that seems not to be my fate. Joseph says he has rubber trees to tend, and the twenty or so people in our boat have crops and hunting and families. We should go. We gather our bags and head down to the longboat. I am comfortably seated when John says softly from the mud bank, "Bill, I think you'd better come up here." The Penan are standing beside the boat. I climb out. La holds out the finished, carved, AK 47 "Tendril" model blow-pipe, holds it out toward me. "It is a gift," says John.

I am stunned, and can say nothing. I take the blowpipe from La's hands slowly, and can only hope that my eyes say it all and that some-day, somehow I can pay him back. We bow and nod and shake, and I climb back in the launch, and we are pushed out, into the current; the engines roar, and the green uncut jungle slides by, and the first rapids spits at us as we head down the Baram, and I head out of Borneo. In the launch, something about the Penan at the riverbank triggers a memory: a passage in Eric Hansen's *Stranger in the Forest*, after he had wandered a year with the Penan:

We discussed acts of violence and crime beyond the jungle. Theft they could understand, but rape, mugging, suicide, and murder were completely foreign to their way of living. Neither one of them could remember a Penan committing any of these acts.

"What would be a serious crime in the Penan community?" I asked.

They conversed for a minute, as though they were having difficulty thinking of any crime. Then Weng explained the concept of *see-hun,* which means to be stingy or not to share.

DOWN, DOWN, dropping Richard at Long Anap, presents and good-byes and good luck, down past Long Akah, where the Chinese owner takes us to the back of his store to see his pet goldfish, a monster in a huge concrete tank, in the shade but under a hole where three roofs pour down fresh rainwater. The fish rises to take food from his master's hand. Down to Long Lama. The deck of the express is full of fresh fish, caught in the Baram by two young boys for sale at Marudi, at least a hundred, several types of catfish, something perchlike, and one ugly creature two and a half feet long and over thirty pounds, looking like a wrestling video parody of a dog salmon, bulging eyes, fat lips, underslung jaw. As we admire the catch another express overtakes us, empty except for four teenage boys; they race us, go in circles around us, and finally ride our wake, the 110-foot boats roaring full speed about four feet apart, boys playing around in the brave new world of machines. Will that list of our crimes, unknown to the Penan, be unintelligible to these boys?

At Marudi, the express is packed. Four days down from Long Moh we pull into Kuala Baram at Miri, past the last ten miles of abandoned logs stacked on both banks, rotting. The wrestling ends and the tape is playing Jimmy Cliff, "By the Waters of Babylon"—by the waters of Babylon, I laid me down and wept.

At the head offices of Samling in Miri, next to the Park Hotel, above the indentured Philippine girls of the Sweet Dreams Beauty Salon and massage parlor, I ask the Samling officers about their operations in relation to native land rights. The answer: "The natives do not have any right to the land. If they or their fathers farmed this or that or hunted here, or their grandfathers were here—about this we do not care. This is an internal problem for the Sarawak government. It is not our concern. We have our obligations and responsibilities. We have the license to remove the timber, and we *must* remove it. It is our duty." The head of field operations echoes: "We operate entirely within the laws of Sarawak. We take no responsibility for those laws."

Finally the weekly flight, direct from Borneo to Tokyo, nonstop. As we approach Narita I wonder how it looked to Jewin Lihan, the rivers below all channeled in concrete, the perfect rice fields, row upon row of plastic-covered greenhouses for the truck farms serving Tokyo.

And in downtown Tokyo, in the Ginza, I walk through two blocks of Mitsubishi buildings to the Soshi Annex, seventh floor. Timber, Paper, Pulp. Workers in identical suits, coats off, are grouped in clusters under signs: Timber Section; North American Team; Tropical Timber Team — a small tribe of timber traders in a clearing of desks. In a private conference room I meet with the charming, obliging, and busy heads of Mitsubishi Tropical Timber. At first I am told that in Borneo the natives benefit, that Penan cannot roam the jungles forever, that blockades are caused by outside agitators. Yet when they realize that I have been there, they stop pushing that drivel; they have lived in Sarawak too, and are not public relations men. Yes, they are getting only four to five tons per acre at Bintulu, taking trees as small as one and a half feet in diameter. Sustainability?

"We would like to do business there for a long time. But five-year licenses? Ten? Fifteen? There's no incentive. The government wants the cutting done fast. If you won't do it, someone else will."

The mood of the room is reflective. I never expected this, but I find myself musing with Mitsubishi. I ask, "What if injustice were proved to you. What if you knew that the Sarawak government was robbing the natives of their land. What would be the response of Mitsubishi?"

They shrug. "We can't do anything in relation to natives. Those are internal problems."

"But without your business, the government would have less incentive to take their land."

"Someone else would do it, someone else would work with the government. Business is business."

THEN THE FLIGHT to America; the plane is one-third full, because Continental Airlines is going broke. In Seattle I am told that half the mills down to Tacoma are now Japanese owned, and that Mitsubishi has just bought the largest timber concession in the history of Canada, up in northern Alberta near Great Slave Lake. They will build a chipping plant. Then the ten-hour drive from Seattle in the winter, across the Cascades, the barren Columbia plateau, the snowy ranges, the empty land. Juliette and I back home after almost two years, standing, at 2 a.m., in the lobby of the Edgewater Motel in our hometown, by a river where people can still catch trout, staring at a handsome, fresh-faced Montana boy behind the desk. "How about a discount?" I say automatically, "We've been away a while." He is filling out our registration form.

"Where to?"

"Asia."

He puts the form on the counter and looks me square in the eye, Montana style:

"How is it down there?"

That stops me. Standing in front of his question, I see again the Penan on the last Marudi-Miri express. The boat is full. Around me, packages and people jam the aisles. After ten minutes of video kung fu I give up my seat and walk back, to go up top and get some air. The last two rows, eight seats, are filled with Penan, bowl haircuts, bracelets, anklets, tattoos. They are all squirming and bright-eyed, watching me. As I approach, one moves over, crowding himself and his child into one seat. He is wearing faded blue shorts, and his white going-to-town shirt has no buttons. His splayed, brown, bare feet are wide and flat on the coke-stained metal floor. Over and over, with short, quick movements he pats the empty seat, fixing me with dark eyes that say, come, come, sit, sit, sit. I sit. He stares at me. "Penan," he says, in English. That's all. Then he motions to the other seven. "Penan," he repeats, happily, serenely, completely, as if words could say no more, as if "Penan" were a vocabulary and syntax, a language, a world. Another whispers "Penan," and all nod and smile vigorously. "Where are you from?" I ask. No contact. I repeat it in Bahasa. Nothing. They look at each other and back to me. Then I open my blue daypack, caked with white mud, in front of their eager eyes, gently pull out my big colored topo map of the Baram, open it section by section—their bouncing increases with each unfolded eagle's view—and spread it on our knees. Eight people crowd around, talking to each other. They are from "Ba Tik"—they say the word all at once—way up in the highlands. The oldest man holds up his hand and indicates walking six days down to Lio Matoh, then four days floating—his hand becomes a canoe in rapids—down the river to the express boat. He bends to the map, and begins to trace the route in reverse, up the Baram, past Long Moh to Lio Matoh, then cross-country up the plateau to the highlands north of Bario. There it would be, though nothing is on the map. "Ba Tik," he says. They all crowd around, hearing the name pronounced again. "Ba Tik," another repeats and all echo. Their hands converge on the map. Eight stubby fingers point the way home.

"It's OK," I say to the boy at the desk.

"Glad to hear it. Welcome home."

UPDATE

It is Sunday, February 6, 1994. We have a Winter Storm Warning for today in western Montana, snow and blowing snow, wind chill of sixty below. It is almost three years since I returned from the tropics to yellow ribbons and a country interested in foreign wars in oil-rich nations, though not so interested in every nation: when I asked a friend in New York publishing how editors might view a book on Borneo natives, she said, " 'Black books don't sell.' That's what they'll say." I mentioned that the Borneo natives are Mongolian, not black. She shrugged. For other reasons, I suppose, the subject should be discomforting: international companies and governments "developing the resources" of people without power . . . that is hard to hear about, because we know if it's happening here, it's happening there, and there, and there.

As this book goes to press, recent developments flood across my modem desk, and the pattern is clear: the good news comes from the international press, passing on news releases from Kuching, the ITTO, the forestry professionals. The bad news comes from the local folks, on the ground, up the Baram. Both sources are accurate.

First, the good news. The Sarawak government has taken the ITTO's advice to slow the cutting. Indeed, the logging is already moving elsewhere. There have been some slumps in the Sarawak timber economy—hard on the State and the natives, although we must remember that the cutting is in poorer hill forest now. At the same time, Kuching has moved toward the Indonesian model of milling more of its own wood, creating jobs in Sarawak instead of exporting so many whole logs to Japan. The United States is beginning to follow suit.

But I am not so interested in the statistics of these articles, news reports, reviews, as in the tone: the assumptions, the context in which the facts are placed. If you begin with government and industry questions—board feet taken per year, sustained yield, environmental consequences, export statistics, value added, jobs—you will come to government and industry answers. Take a professional and vaguely "liberal" source as an example of this reasonable tone, a review in *Conservation Biology* in March 1991 of three books on Sarawak: a World Wildlife Fund book, the ITTO report, and the U.S. Congressional Mission report of 1989. The knowledgeable reviewer dismisses the slanted and inaccurate congressional report, and is skeptical of the ITTO; what's interesting is his summary of the report from the environmentally conscious World Wildlife Fund. We might call this a liberal review of liberals. How then can it serve the timber interests? The reviewer says: "Logging concessions are not granted on NCR (Native Customary Rights) land, but often are on nearby forested land that is legally owned by the government but used and frequently claimed by the villagers. Logging companies must legally compensate local villagers. . . . In practice, logging companies usually make unofficial payments to villagers even when the land is not legally held under NCR . . . in order to obtain the goodwill of the villagers and avoid trouble later. . . . If the camp managers refuse to pay the claims, certain villagers may get so angry that they try to blockade a road to get the compensation they believe they are entitled to or to protest the further logging of the forest." Ignoring minor inaccuracies (companies do not have to compensate), we might think this sounds pretty fair. The rhetoric comes from Kuching: what is "legally owned" is associated with the government, and attempts to "obtain goodwill" are associated with timber companies, whereas "anger" is the property of villagers. In fact, "certain villagers," echoing government theories of instigators and troublemakers, are "so angry" that they blockade "to get" compensation.

Later the reviewer notes that in a longhouse which the congressional delegation visited with environmentalists, "20 of the 70 villagers interviewed favored the continued commercial logging. . . . Presumably in other villages the majority of people would actually favor logging." The reviewer seems not to know that most headmen support logging and are bribed, that a third of most longhouses are related to the headman by family or politics, and that for anyone to counter the headman's decisions is to break *adat*, the customary harmony/law of the longhouse. Therefore a 50 to 20 opposition voiced to outsiders is remarkable, indeed revolutionary; moreover, I found such ratios to be typical all the way up the Baram.

The above is a bit like discussing slavery in 1860 by accepting it as the law of the land, and then discussing the need for mitigation, such as greater compensation to slaves. Although the reviewer knows that NCR land is only the cleared land, and not the hunted forest, and for that matter mainly the land cleared before 1958, he grants the government legal ownership of the rest of the forest—which until thirty years ago was used exclusively by these hunting-gathering societies and has only recently been taken away ("There are no native land rights"). In this context, natives are perceived as protesting in order "to get the compensation." In fact such compensation, while continuing present practices (we used to call this Uncle Tomism), is the main conclusion of the review: "In summary, all three reports agree that the government should give greater consideration to the welfare of native people *in granting logging rights to timber companies.* The three reports also agree that most natives involved in the blockades of timber roads are predominantly concerned with receiving what they believe to be fair compensation. . . . *Only a small percentage of the natives, mostly Penans, actually want to stop the logging. . . .*" (Emphases mine.) Actually, almost all natives, including headmen, would like *full control* of the logging in their district, right now.

What if we discussed activism among the Blackfeet Indians in Montana without recognizing that their present (legal) reservation is about one-tenth of their previous territory (which included Glacier Park), that modernization has been forced on them, and that systematic attempts have been made to destroy their culture and ecosystem. Might their unrest seem simply a way to increase welfare payments? And then might the answer be to toss a few more coins?

In the *Conservation Biology* review, the writer knows the government policies, the forest data, and the forest, but he does not know what is happening upriver. Have no doubt: the Sarawak and especially the Malaysian governments are quite progressive; the Sarawak government is indeed slowing the logging from three times the sustainable yield to twice the sustainable yield, and they are genuinely interested in some biodiversity and park arrangements.

But this set of questions and forestry statistics ignores, or puts to the side, native land rights, economic coercion, fundamental injustice, as if those were issues marginal to getting the timber out.

That is why, at a conference on Malaysia and the environment at

Ohio University in April 1993, with leading Malaysian ministers, foresters, and environmentalists, I found myself in a strange position. I had feared, as a scholar, that I might have been duped (as the Sarawak ministers claimed), that suddenly my understanding would be proved wrong. I was nervous as I read to a room of Malaysian experts a paper on Samling's bribing, cutting, and compensation in the middle Baram. It turned out, however, that I was the only one in the room who had ever been up the Baram River, though one Malay had been to Marudi, and one American Penan expert had been to Mulu. No one at the conference had the faintest idea what was happening between local timber companies and the longhouses—or rather they all suspected, and didn't want to think about it. Participants were delighted to talk of environmentalism, water quality, the Rio summit, forestry, statistics, all of which amount to fine-tuning the progressive nation-state. However, no one wanted to hear of native land rights. That would sink the entire enterprise. Finally, late one night, I was with a Malaysian official who had been my opponent, and we were talking privately. I was still wondering if I had misunderstood something fundamental when he said casually, "Of course if I were a native in the longhouse, I'd do it." "Do what?" "I'd blockade the logging roads. What else can they do?"

So what is the news, from upriver, on the ground? Remember Long Terawan, up the Tutoh on the way to Mulu Park, with the Chinese storeboat moored at the dock and the old teacher who had met the last Rajah Brooke? The natives there are Berawan, and now they are protesting the Royal Mulu Resort, a five-star hotel (and planned golf course) built near the park by the Japanese chain Rihga Royal Hotels, on fifty acres (so far) of native land. In the skirmish some natives have suspended services such as guiding, and delivering fuel to the hotel. Interestingly, the natives are also trying to defend their own fledgling tourist services to the park (itself carved from ancestral Penan and Berawan land). The government, however, while arguing that natives must be brought into the modern world, has never encouraged or protected native businesses trying to capitalize on modernization, in this case their hunting ground evolving into a tourist attraction. The native protests were covered in the Sarawak and Kuala Lumpur press as actions of "a certain group inciting the Berawan community to demand for excessive compensation."

But beyond the sad and predictable stories — Iban up the Tinjar protesting gigantic new oil palm plantations in their cutover district —
there is one incident that particularly troubles me. Remember the bulldozer, covered with vines and rusting at the end of the road in the
upper Baram? And La (not his real name), the Penan from Ba Muboi
who had stopped the surveyor and the bulldozer, who ate my choco-
lates, who told of the children pouring Mr. Sei's candy on the ground,
who gave me the carved AK 47 blowpipe now hanging above my computer?

In 1992, Samling tore away the vines, cleaned up the bulldozer,
and started pushing the road farther up the Baram, without a timber
agreement. In January 1993, the Penan of Ba Muboi and Ba Ajeng on
the Salaan River in the bulldozer's path, with over a thousand Penan,
blocked the logging road. (In some updates and Internet releases these
communities are called Long Ajang and Long Mobui; "Ba" is Penan
for "river." "Muboi" is the spelling I was given there, and the one used
by Thomas Jalong in correspondence.) On September 28, 1993, after
nine months of blockade, police (apparently many, with riot gear) were
trucked up the new road, supported by a helicopter. They forcefully
broke the blockade using tear gas and batons. Reports vary, but up
to thirty-nine natives were arrested and jailed. A month later, on the
second of November, three truckloads of "army" came to Ba Muboi,
where Mr. Sei's candy and chainsaw diplomacy had failed, demanded
to know the whereabouts of two escaped leaders, and threatened to
bomb the village if inhabitants did not cooperate. The military report-
edly offered bounties for the escaped leaders, dead or alive. Isn't that
headhunting?

As for La — is he one of those arrested and jailed? One of the escaped leaders, with a price on his head? I do not know — and if I did,
I would not tell.

As of June 1994, the Muboi blockades continued, the police actions
and violence were escalating, and Samling was still advancing up the
Baram. In that district, where there is a profit to be made, there is
no 30 percent slowing of the cutting — only of the cut in poorer and
steeper hill forest. The statistics, then, reflect lower yield (a forestry
point of view), yet Samling is still moving through the natives' land at
fifty acres an hour (a native point of view).

Has the upper Baram signed a timber agreement? In answer to my questions, Thomas Jalong wrote from the SAM office in Marudi: "Despite the strong objection made by the people of Long Salaan and Long Moh during the said meeting at Long Moh [Chapter 12], the company has not given up. Some months later, the company picked up few representatives from Long Salaan and brought them to Miri where they can apple-polish the representatives. The same technique also applied to Long Moh. . . . Representatives brought down to Miri usually put in Cosy Hotels and given food and good amount of allowances . . . the representatives usually do not understand what they are signing as they are illiterate." After that letter, during 1994, the system slouched to its conclusion; the headmen of Long Salaan and Long Moh, with some committee members, signed an agreement at Miri, without longhouse ratification. Were the headmen kicked out of the longhouse, as promised in that heated meeting of 1990? Apparently not. With some sympathy, I can imagine them facing the familiar Samling arguments, beyond the bribes: you can't stop us, you have no rights, this package is the best you can get for your people, take it or leave it. In newsprint around the world this emerges as "compensation."

Finally, I asked Thomas about our friends in Long Anap, down in the middle Baram, and Fujino, in the timber camp on the ridge above. In 1993, the news was good: "The area defended by the people of Long Anap is still intact but presently the company is trying to force its way to log the area. We'll see what happens next." For three years Long Anap had blockaded and resisted; the 1991 coupe was uncut. I know of no other instance in Sarawak where one longhouse had actually stopped the timber company for so long. It must have been quite a battle of nerves between Long Anap and Fujino.

Six months later, however, in February 1994, the news was different: "The forest area that the people of Long Anap have been protecting is partially logged out. . . . they were told that the government can allocate only five acres per door [family] as communal forest reserves." An acre per person, of course, will not support a hunting-gathering ecology. Anap has lost, though apparently they have wrung some unusual "compensation" from the government.

The surprising news is that in 1993 Samling transferred Fujino to Papua New Guinea. Some other manager was "trying to force" his way in. Was Fujino too kind to Long Anap? Is that why they could hold out for so long? Was he transferred for promotion, or punishment? Or simply because Samling sees the end in the Baram, and needs its most experienced people elsewhere?

Which brings us to the end of this story. The Sarawak cut is declining, but there is little indication that natives will emerge with more political control over their lives. That is, almost all recommended environmental measures could be enacted, and yet the control of the forest would still have shifted from the natives who live there and have used those resources for centuries, to the government and its allied entrepreneurs, almost all Malay and Chinese.

Already the action is moving on—logging in Papua New Guinea increased 400 percent from 1992 to 1993. How do you think the natives there are faring? Mitsubishi is half owner (with its trading partner) of the Amazon's biggest plywood operation, Eidai. And sure enough, on Internet as we go to press: Samling Timber, through a consortium called Berama Company Limited, has signed a fifty-year contract with the government in Guyana, South America, to log over a million and a half hectares of rainforest: "Their concession overlaps the lands of some 1,200 Amerindians, most of whom lack clear land titles. They are demanding that their land rights be secured before more logging takes place." Until Russia straightens itself out, Siberia is unpredictable.

What can be done? Almost a third of Sarawak is still unlogged. There will be other Long Anaps and Ba Mubois needing encouragement in remote hills at the top of the Baram. Land use policies can be rewritten, though as I hope to have made clear, Malaysians may easily perceive first world interference—neocolonialism—where we perceive a legitimate interest in human and land rights.

As I have lectured and shown slides on this subject over the last three years, people invariably ask what action they might take. One could support revival of the BTU tax proposal or some equivalent taxing of all energy consumption in America, for first world consumption underwrites resource extraction around the world. In the current debate over American forests, we still lack accurate projections of long-term sustained yield, and realistic pricing. We can pressure Japan to respect indigenous peoples in their enterprises abroad, and we can ask the U.S. government also to conceive of human rights in relation to commerce as well as politics.

Mainly, whatever our individual efforts might be, we can encourage vigilance, for colonialism is not dead: as the worldwide race for dwindling resources quickens, rapid resource extraction by the few is often at the expense of the many, the indigenous, and the children of us all.

NOTES

CHAPTER 1. INTO THE HEART

The 1988 timber agreement between Samling Timber Company and the middle Baram will be discussed in more detail in Chapter 10. The one-page summary of native land rights which Fujino read, and which survived the readings of three lawyers of different persuasions and two government ministers, is printed on page 16.

CHAPTER 2. GOING TO TOWN

To pursue the comparison to Fort Benton, readers might look at the Montana Centennial anthology, *The Last Best Place,* edited by Kittredge and Smith, especially the selections from Fort Benton journals and Captain Mullan, and the up-river descriptions by Catlin, Audubon, Harris, and Point in the 1830s and 1840s.

Sahabat Alam Malaysia is the Malaysian affiliate of Friends of the Earth. The headquarters is in Penang, Malaysia, on the mainland. Marudi hosts the field office in Sarawak, founded by Evelyne Hong and Harrison Ngau in 1982.

CHAPTER 3. RAJAHS AND RANEES

Sylvia Brett's writings are from her autobiography, *Queen of the Head Hunters;* the author is listed as Sylvia, Lady Brooke. Tom Harrisson's assessment of dragon jars and human life appears in his account of the Bario highlands before and during World War II, *World Within.*

A useful introduction to the history of the period is Robert Payne's *The White Rajahs of Sarawak,* which is quoted on pages 44, 46, 47, and 50. The "Brooke officers" are quoted in Charles Allen's *Tales from the South China Seas* (chapters 2 and 6), a collection of oral history accounts and remembrances of British rule in the area.

Ranee Margaret Brooke's tribute to Charles appears in her memoir, *My Life in Sarawak.*

CHAPTER 4. UPRIVER

I discussed the demise of the buffalo in chapter 4 of a previous work, *Ten Tough Trips: Montana Writers and the West.* The quotations from Sheridan, Alderson, and various historical sources appear in that chapter. The quotation from McNickle is from chapter 6.

CHAPTER 5. TREE, FOREST, LOG

The literature of rainforest ecology is vast and growing. I read most of what was available in English on the Sarawak forest and will not list individual articles, although some authors and presses deserve mention: Vinson Sutlive and Vandana Shiva have done valuable work, and the various publications of SAM, the Third World Network, and the Consumers Association of Penang print or reprint the bulk of environmental and nongovernmental views on Southeast Asian forests.

Some major books on the Sarawak forest or related issues include: Hanbury-Tenison, *Mulu, the Rainforest;* Wallace, *The Malay Archipelago;* Bethel et al., *The Role of U.S. Multinational Corporations in Commercial Forestry Operations in the Tropics;* Richards, *The Tropical Rainforest;* World Bank, *Tribal Peoples and Economic Development;* Forestry Department of Malaysia, *Forestry in Malaysia* (1988); Hurst, *Rainforest Politics;* Hong, *Natives of Sarawak;* and Caufield, *In the Rainforest.* In summarizing the ecology of the rainforest, I have tried to follow or at least not contradict Caufield's book, itself a compendium of rainforest research up to 1985. That gives us a single source, first published in the *New Yorker,* with which to begin debate.

CHAPTER 6. SUNRISE IN JAPAN

Apart from the well-known "Japan bashing" books of the 1980s (Johnson, Prestowitz, Fallows, van Wolferen—only the last really bashes), readers might be interested in Kenichi Ohmae's *Fact and Friction.* He is a respected Japanese businessman, economist, physicist (MIT Ph.D.), and columnist on Japan-American trade relations.

Out of the scores of "Japan observing" books one can find, I would call attention to Morley's *Pictures from the Water Trade,* Whiting's *You Gotta Have Wa,* Field's *In the Realm of the Dying Emperor,* Reischauer's *The Japanese Today,* and Smith's *Japanese Society.* There are also over fifty modern (Meiji to present) Japanese works of literature in print in English, many of them excellent.

Kuroda's book, co-authored with François Nectoux and available in English through the World Wildlife Fund, is *Timber from the South Seas: An Analysis of Japan's Tropical Timber Trade and Its Environmental Impact* (1989). Research on the Philippines, Smithsonian calculations, and the timber trader's comments appear in Kuroda. Any other unattributed quotations in the chapter are also from Kuroda.

CHAPTER 7. TOKYO RISING

The wood use data are from Kuroda. The Tokyo motor show was reported in the *Japan Times.* The Shinjuku office building statistics are from Tokyo newspapers and the staff of JATAN, Tokyo.

CHAPTER 8. JAMES WONG, JAPAN, AND THE 1987 BLOCKADES

James Wong's warm and graceful memoir and compilation of his father's letters is called *The World According to William Wong Tsap En, 'No Joke James.'* He also wrote a history of his political imprisonment under the same security act that netted Harrison Ngau, entitled *The Price of Loyalty.* He has also made his views known in numerous interviews, columns, features, letters to the editor, and other publications. Wong replied to an Australian environmental piece called "Rumble in the Jungle" with a five-part article called "Stumble in the Jungle" published in Sarawak's *People's Mirror* (April 11–15, 1988); in 1991 he offered enlarged, bound copies to visitors to his Ministry of Tourism and Environment at the State Building in Kuching.

The history of the Chinese in Sarawak is from Daniel Chew's *Chinese Pioneers on the Sarawak Frontier, 1841–1941,* part of a fine series of South-East Asian Historical Monographs from Oxford University Press in Singapore. Wong's description of entering trading and then timber after 1946 is from chapter 1 of *The Price of Loyalty.*

The reports of the Stampin blockade by a CAP reporter appeared in the *Utusan Konsumer* in March 1987.

CHAPTER 9. BRUNO MANSER AND THE PENAN

Wallace's *The Malay Archipelago* and Brooke's journals, some quoted in Payne, sketch the scenes of their meetings, and Redmond O'Hanlon tells of Lord Monboddo's ape in *Into the Heart of Borneo.* Beccari is quoted and dissected by Harrisson in *Borneo Jungle,* his account of the Oxford University Expedition up the Baram in 1932. The account of men "in a state of nature" carrying mates away into the jungle is quoted by O'Hanlon in his chapter on Darwin and Wallace (chap. 13). His book on *Conrad and Darwin* is also excellent.

The brief account of Manser's background in Switzerland is from private correspondence between Manser and myself; Davis's account of Manser in Sarawak appeared in *Outside,* January 1991.

The ethnographic and botanical accounts of Penan life, when not attributed to Davis, are from the *Sarawak Museum Journal,* Special Issue 4, part 3, the Orang Ulu Seminar (December 1989). Details of Manser in Sarawak are as told by Manser to Davis or myself, unless a newspaper is cited. The pages from Manser's journals, with illustrations, were kindly sent by Manser. The quotations from Manser in Tokyo are transcribed from my videotape. Thom Henley, and Randy Hayes of Rainforest Action Network, are quoted by Davis in *Outside.*

CHAPTER 10. LONG ANAP

All of the descriptions of life in Long Anap are from personal observations and from informants who were there with me at the time. A complete copy of Samling's 1988 middle Baram timber agreement is in my possession.

CHAPTER 11. TIMBER CAMP

Much of the timber industry literature consists of reprints of newspaper articles from Sarawak and Japan. The ITTO reports are public.

CHAPTER 12. MEETING AT LONG MOH

The sources for the economic summaries are myself, Joseph Wang Tingang, the literature of various timber, environmental, and opposition organizations, natives in the longhouses affected, and what I learned at Fujino's timber camp. These diverse sources agree quite closely on the amount of the cut, its variance from sustained yield, and compensation to natives. There are discrepancies concerning the gross value of the timber (what percentage is meranti?) and, with operating costs kept private, the net profit.

SELECTED BIBLIOGRAPHY

Allen, Charles, ed. *Tales from the South China Seas*. 1983; rpt. London: Futura, 1984.

Anderson, Robert S., and Walter Huber. *The Hour of the Fox: Tropical Forests, the World Bank, and Indigenous People in Central India*. Seattle: University of Washington Press, 1988.

Belcher, Martha, and Angela Gennino. *Southeast Asia Rainforests: A Resource Guide and Directory*. San Francisco: Rainforest Action Network, 1993.

Bethel, James S., et al. *The Role of U.S. Multinational Corporations in Commercial Forestry Operations in the Tropics*. Seattle: College of Forest Resources, University of Washington, 1982.

Bevis, William W. *Ten Tough Trips: Montana Writers and the West*. Seattle: University of Washington Press, 1990.

Broad, Robin, and John Cavanagh. *Plundering Paradise: The Struggle for the Environment in the Philippines*. Berkeley: University of California Press, 1993.

Brooke, Lady Margaret. *My Life in Sarawak*. 1913; rpt. Singapore: Oxford University Press, 1986.

Brooke, Lady Sylvia. *Queen of the Head Hunters*. New York: Morrow, 1972; rpt. Singapore: Oxford University Press, 1990.

Caufield, Catherine. *In the Rainforest*. New York: Alfred A. Knopf, 1984; rpt. Chicago: University of Chicago Press, 1986.

Chew, Daniel. *Chinese Pioneers on the Sarawak Frontier, 1841–1941*. Singapore: Oxford University Press, 1990.

Chiang, Lim Poh. *Among the Dyaks*. Singapore: Graham Brash, 1989.

Colchester, Marcus. *Pirates, Squatters and Poachers: The Political Ecology of Dispossession of the Native Peoples of Sarawak*. London: Survival International and INSAN, 1989.

Davis, Shelton. *Victims of the Miracle*. Cambridge: Cambridge University Press, 1977.

Davis, Wade. "The Apostle of Borneo." *Outside* Magazine, January 1991, pp. 31ff.

Davis, Wade, and Thom Henley. *Penan: Voice for the Borneo Rainforest*. Vancouver: Western Canada Wilderness Committee, 1990.

Fallows, James. *More Like Us*. Boston: Houghton Mifflin, 1989.

Field, Norma. *In the Realm of the Dying Emperor*. New York: Pantheon, 1991.

Forestry Department of Malaysia. *Forestry in Malaysia*. Kuala Lumpur: Forestry Department of Malaysia, 1988.

Furtado, J. I., ed. *Tropical Ecology and Development*. Kuala Lumpur: International Society of Tropical Ecology, 1980.

Hanbury-Tenison, Robin. *Mulu, the Rainforest.* London: Weidenfeld and Nicolson, 1980.

Hansen, Eric. *Stranger in the Forest: On Foot Across Borneo.* Boston: Houghton Mifflin, 1988; rpt. New York: Penguin, 1989.

Harrisson, Tom. *Borneo Jungle.* 1938; rpt. Singapore: Oxford University Press, 1988.

————. *World Within: A Borneo Story.* London: Cresset Press, 1959; rpt. Singapore: Oxford University Press, 1986.

Hong, Evelyne. *Natives of Sarawak.* Penang: Institut Masyarakat, 1987.

Hose, Charles. *The Field-Book of a Jungle Wallah.* 1929; rpt. Singapore: Oxford University Press, 1985.

Hurst, Philip. *Rainforest Politics: Ecological Destruction in South-East Asia.* London and Atlantic Highlands, N.J.: Zed Books, 1990.

Idris, S. M. Mohd., ed. *The Battle for Sarawak's Forests.* 2d edition. Penang: World Rainforest Movement and Sahabat Alam Malaysia, 1990.

International Tropical Timber Organization. *The Promotion of Sustainable Forest Management: A Case Study in Sarawak, Malaysia.* Yokohama: ITTO, 1990.

Johnson, Chalmers. *MITI and the Japanese Miracle: The Growth of Industrial Policy, 1925–1975.* Stanford: Stanford University Press, 1982; rpt. Tokyo: Charles E. Tuttle, 1986.

Kavanagh, M., A. A. Rahim, and J. C. Hails. *Rainforest Conservation in Sarawak.* Kuala Lumpur: World Wildlife Fund Malaysia, 1989.

Keith, Agnes Newton. *Three Came Home.* 1948; rpt. Pataling Jaya: Eastview Productions, 1982.

Kittredge, William, and Annick Smith, eds. *The Last Best Place.* Helena: Montana Historical Society, 1988.

Linklater, Andro. *Wild People: Travels with Borneo's Head-Hunters.* New York: Atlantic Monthly Press, 1990.

Manning, Richard. *The Last Stand.* Salt Lake City: Peregrine Smith, 1991.

Morley, John David. *Pictures from the Water Trade.* New York: Atlantic Monthly Press, 1985.

Nectoux, François, and Yoichi Kuroda. *Timber from the South Seas: An Analysis of Japan's Tropical Timber Trade and Its Environmental Impact.* Gland, Switzerland: World Wildlife Fund International, 1989.

O'Hanlon, Redmond. *Into the Heart of Borneo.* New York: Random House, 1984; rpt. New York: Random House Vintage, 1987.

Ohmae, Kenichi. *Fact and Friction.* Tokyo: The Japan Times Ltd., 1990.

Payne, Robert. *The White Rajahs of Sarawak.* New York: Funk and Wagnalls, 1960; rpt. Singapore: Oxford University Press, 1986.

Prestowitz, Clyde V. *Trading Places: How America Allowed Japan to Take the Lead.* Tokyo: Charles E. Tuttle, 1988.

Raghavan, Chakravarthi. *Recolonization: GATT, the Uruguay Round and the Third World.* Penang: Third World Network, 1990.

Reischauer, Edwin O. *The Japanese Today: Changes and Continuity.* Cambridge: Harvard University Press, 1988.

Richards, Paul. *The Tropical Rain Forest.* Cambridge: Cambridge University Press, 1981.

Sarawak Museum Journal. Special Issue 4, part 3, *Orang Ulu Cultural Heritage Seminar.* Kuching: Sarawak Museum, 1989.

Sesser, Stan. "Logging the Rain Forest." *New Yorker,* May 27, 1991, pp. 42–67.

Shiva, Vandana. *Forestry Crisis and Forestry Myths.* Penang: World Rainforest Movement, 1987.

Smith, Robert J. *Japanese Society: Tradition, Self and the Social Order.* Cambridge and New York: Cambridge University Press, 1983.

Sutlive, Vinson H., et al., eds. *Blowing in the Wind: Deforestation and Long-range Implications.* Williamsburg: Department of Anthropology, College of William and Mary, 1981.

Tate, D. J. M. *Rajah Brooke's Borneo.* Hong Kong: John Nicholson, 1988.

Tsing, Anna Lowenhaupt. *In the Realm of the Diamond Queen: Marginality in an Out-of-the-Way Place.* Princeton: Princeton University Press, 1993.

U.S. Congressional Staff Study Mission to Malaysia. *The Tropical Timber Industry in Sarawak, Malaysia.* Washington, D.C.: U.S. Government Printing Office, 1989.

Wallace, Alfred Russel. *The Malay Archipelago.* 1869; rpt. Singapore: Oxford University Press, 1989.

Whiting, Robert. *You Gotta Have Wa.* New York: Macmillan, 1989.

Wolferen, Karel van. *The Enigma of Japanese Power: People and Politics in a Stateless Nation.* New York: Knopf, 1989.

Wong, James. *The Price of Loyalty.* Singapore: Summer Times Publishing, 1983.

Wong, James, ed. *The World According to William Wong Tsap En: 'No Joke, James'.* Singapore: Summer Times Publishing, 1985.

World Bank. *Tribal Peoples and Economic Development: Human Ecological Considerations.* Washington, D.C.: World Bank, 1982.